Mometrix
TEST PREPARATION

Virginia SOL Grade 5 Science

Secrets Study Guide

AF834555

Dear Future Exam Success Story

First of all, **THANK YOU** for purchasing Mometrix study materials!

Second, congratulations! You are one of the few determined test-takers who are committed to doing whatever it takes to excel on your exam. **You have come to the right place.** We developed these study materials with one goal in mind: to deliver you the information you need in a format that's concise and easy to use.

In addition to optimizing your guide for the content of the test, we've outlined our recommended steps for breaking down the preparation process into small, attainable goals so you can make sure you stay on track.

We've also analyzed the entire test-taking process, identifying the most common pitfalls and showing how you can overcome them and be ready for any curveball the test throws you.

Standardized testing is one of the biggest obstacles on your road to success, which only increases the importance of doing well in the high-pressure, high-stakes environment of test day. Your results on this test could have a significant impact on your future, and this guide provides the information and practical advice to help you achieve your full potential on test day.

Your success is our success

We would love to hear from you! If you would like to share the story of your exam success or if you have any questions or comments in regard to our products, please contact us at **800-673-8175** or **support@mometrix.com**.

Thanks again for your business and we wish you continued success!

Sincerely,
The Mometrix Test Preparation Team

Need more help? Check out our flashcards at:
http://MometrixFlashcards.com/SOL

Written and edited by the Mometrix Exam Secrets Test Prep Team
Printed in the United States of America

TABLE OF CONTENTS

Introduction

Thank you for purchasing this resource! You have made the choice to prepare yourself for a test that could have a huge impact on your future, and this guide is designed to help you be fully ready for test day. Obviously, it's important to have a solid understanding of the test material, but you also need to be prepared for the unique environment and stressors of the test, so that you can perform to the best of your abilities.

For this purpose, the first section that appears in this guide is the **Secret Keys**. We've devoted countless hours to meticulously researching what works and what doesn't, and we've boiled down our findings to the five most impactful steps you can take to improve your performance on the test. We start at the beginning with study planning and move through the preparation process, all the way to the testing strategies that will help you get the most out of what you know when you're finally sitting in front of the test.

We recommend that you start preparing for your test as far in advance as possible. However, if you've bought this guide as a last-minute study resource and only have a few days before your test, we recommend that you skip over the first two Secret Keys since they address a long-term study plan.

If you struggle with **test anxiety**, we strongly encourage you to check out our recommendations for how you can overcome it. Test anxiety is a formidable foe, but it can be beaten, and we want to make sure you have the tools you need to defeat it.

Secret Key #1 – Plan Big, Study Small

There's a lot riding on your performance. If you want to ace this test, you're going to need to keep your skills sharp and the material fresh in your mind. You need a plan that lets you review everything you need to know while still fitting in your schedule. We'll break this strategy down into three categories.

Information Organization

Start with the information you already have: the official test outline. From this, you can make a complete list of all the concepts you need to cover before the test. Organize these concepts into groups that can be studied together, and create a list of any related vocabulary you need to learn so you can brush up on any difficult terms. You'll want to keep this vocabulary list handy once you actually start studying since you may need to add to it along the way.

Time Management

Once you have your set of study concepts, decide how to spread them out over the time you have left before the test. Break your study plan into small, clear goals so you have a manageable task for each day and know exactly what you're doing. Then just focus on one small step at a time. When you manage your time this way, you don't need to spend hours at a time studying. Studying a small block of content for a short period each day helps you retain information better and avoid stressing over how much you have left to do. You can relax knowing that you have a plan to cover everything in time. In order for this strategy to be effective though, you have to start studying early and stick to your schedule. Avoid the exhaustion and futility that comes from last-minute cramming!

Study Environment

The environment you study in has a big impact on your learning. Studying in a coffee shop, while probably more enjoyable, is not likely to be as fruitful as studying in a quiet room. It's important to keep distractions to a minimum. You're only planning to study for a short block of time, so make the most of it. Don't pause to check your phone or get up to find a snack. It's also important to **avoid multitasking**. Research has consistently shown that multitasking will make your studying dramatically less effective. Your study area should also be comfortable and well-lit so you don't have the distraction of straining your eyes or sitting on an uncomfortable chair.

The time of day you study is also important. You want to be rested and alert. Don't wait until just before bedtime. Study when you'll be most likely to comprehend and remember. Even better, if you know what time of day your test will be, set that time aside for study. That way your brain will be used to working on that subject at that specific time and you'll have a better chance of recalling information.

Finally, it can be helpful to team up with others who are studying for the same test. Your actual studying should be done in as isolated an environment as possible, but the work of organizing the information and setting up the study plan can be divided up. In between study sessions, you can discuss with your teammates the concepts that you're all studying and quiz each other on the details. Just be sure that your teammates are as serious about the test as you are. If you find that your study time is being replaced with social time, you might need to find a new team.

Secret Key #2 – Make Your Studying Count

You're devoting a lot of time and effort to preparing for this test, so you want to be absolutely certain it will pay off. This means doing more than just reading the content and hoping you can remember it on test day. It's important to make every minute of study count. There are two main areas you can focus on to make your studying count.

Retention

It doesn't matter how much time you study if you can't remember the material. You need to make sure you are retaining the concepts. To check your retention of the information you're learning, try recalling it at later times with minimal prompting. Try carrying around flashcards and glance at one or two from time to time or ask a friend who's also studying for the test to quiz you.

To enhance your retention, look for ways to put the information into practice so that you can apply it rather than simply recalling it. If you're using the information in practical ways, it will be much easier to remember. Similarly, it helps to solidify a concept in your mind if you're not only reading it to yourself but also explaining it to someone else. Ask a friend to let you teach them about a concept you're a little shaky on (or speak aloud to an imaginary audience if necessary). As you try to summarize, define, give examples, and answer your friend's questions, you'll understand the concepts better and they will stay with you longer. Finally, step back for a big picture view and ask yourself how each piece of information fits with the whole subject. When you link the different concepts together and see them working together as a whole, it's easier to remember the individual components.

Finally, practice showing your work on any multi-step problems, even if you're just studying. Writing out each step you take to solve a problem will help solidify the process in your mind, and you'll be more likely to remember it during the test.

Modality

Modality simply refers to the means or method by which you study. Choosing a study modality that fits your own individual learning style is crucial. No two people learn best in exactly the same way, so it's important to know your strengths and use them to your advantage.

For example, if you learn best by visualization, focus on visualizing a concept in your mind and draw an image or a diagram. Try color-coding your notes, illustrating them, or creating symbols that will trigger your mind to recall a learned concept. If you learn best by hearing or discussing information, find a study partner who learns the same way or read aloud to yourself. Think about how to put the information in your own words. Imagine that you are giving a lecture on the topic and record yourself so you can listen to it later.

For any learning style, flashcards can be helpful. Organize the information so you can take advantage of spare moments to review. Underline key words or phrases. Use different colors for different categories. Mnemonic devices (such as creating a short list in which every item starts with the same letter) can also help with retention. Find what works best for you and use it to store the information in your mind most effectively and easily.

Secret Key #3 – Practice the Right Way

Your success on test day depends not only on how many hours you put into preparing, but also on whether you prepared the right way. It's good to check along the way to see if your studying is paying off. One of the most effective ways to do this is by taking practice tests to evaluate your progress. Practice tests are useful because they show exactly where you need to improve. Every time you take a practice test, pay special attention to these three groups of questions:

- The questions you got wrong
- The questions you had to guess on, even if you guessed right
- The questions you found difficult or slow to work through

This will show you exactly what your weak areas are, and where you need to devote more study time. Ask yourself why each of these questions gave you trouble. Was it because you didn't understand the material? Was it because you didn't remember the vocabulary? Do you need more repetitions on this type of question to build speed and confidence? Dig into those questions and figure out how you can strengthen your weak areas as you go back to review the material.

Additionally, many practice tests have a section explaining the answer choices. It can be tempting to read the explanation and think that you now have a good understanding of the concept. However, an explanation likely only covers part of the question's broader context. Even if the explanation makes perfect sense, **go back and investigate** every concept related to the question until you're positive you have a thorough understanding.

As you go along, keep in mind that the practice test is just that: practice. Memorizing these questions and answers will not be very helpful on the actual test because it is unlikely to have any of the same exact questions. If you only know the right answers to the sample questions, you won't be prepared for the real thing. **Study the concepts** until you understand them fully, and then you'll be able to answer any question that shows up on the test.

It's important to wait on the practice tests until you're ready. If you take a test on your first day of study, you may be overwhelmed by the amount of material covered and how much you need to learn. Work up to it gradually.

On test day, you'll need to be prepared for answering questions, managing your time, and using the test-taking strategies you've learned. It's a lot to balance, like a mental marathon that will have a big impact on your future. Like training for a marathon, you'll need to start slowly and work your way up. When test day arrives, you'll be ready.

Start with the strategies you've read in the first two Secret Keys—plan your course and study in the way that works best for you. If you have time, consider using multiple study resources to get different approaches to the same concepts. It can be helpful to see difficult concepts from more than one angle. Then find a good source for practice tests. Many times, the test website will suggest potential study resources or provide sample tests.

Practice Test Strategy

If you're able to find at least three practice tests, we recommend this strategy:

Untimed and Open-Book Practice

Take the first test with no time constraints and with your notes and study guide handy. Take your time and focus on applying the strategies you've learned.

Timed and Open-Book Practice

Take the second practice test open-book as well, but set a timer and practice pacing yourself to finish in time.

Timed and Closed-Book Practice

Take any other practice tests as if it were test day. Set a timer and put away your study materials. Sit at a table or desk in a quiet room, imagine yourself at the testing center, and answer questions as quickly and accurately as possible.

Keep repeating timed and closed-book tests on a regular basis until you run out of practice tests or it's time for the actual test. Your mind will be ready for the schedule and stress of test day, and you'll be able to focus on recalling the material you've learned.

Secret Key #4 – Pace Yourself

Once you're fully prepared for the material on the test, your biggest challenge on test day will be managing your time. Just knowing that the clock is ticking can make you panic even if you have plenty of time left. Work on pacing yourself so you can build confidence against the time constraints of the exam. Pacing is a difficult skill to master, especially in a high-pressure environment, so **practice is vital**.

Set time expectations for your pace based on how much time is available. For example, if a section has 60 questions and the time limit is 30 minutes, you know you have to average 30 seconds or less per question in order to answer them all. Although 30 seconds is the hard limit, set 25 seconds per question as your goal, so you reserve extra time to spend on harder questions. When you budget extra time for the harder questions, you no longer have any reason to stress when those questions take longer to answer.

Don't let this time expectation distract you from working through the test at a calm, steady pace, but keep it in mind so you don't spend too much time on any one question. Recognize that taking extra time on one question you don't understand may keep you from answering two that you do understand later in the test. If your time limit for a question is up and you're still not sure of the answer, mark it and move on, and come back to it later if the time and the test format allow. If the testing format doesn't allow you to return to earlier questions, just make an educated guess; then put it out of your mind and move on.

On the easier questions, be careful not to rush. It may seem wise to hurry through them so you have more time for the challenging ones, but it's not worth missing one if you know the concept and just didn't take the time to read the question fully. Work efficiently but make sure you understand the question and have looked at all of the answer choices, since more than one may seem right at first.

Even if you're paying attention to the time, you may find yourself a little behind at some point. You should speed up to get back on track, but do so wisely. Don't panic; just take a few seconds less on each question until you're caught up. Don't guess without thinking, but do look through the answer choices and eliminate any you know are wrong. If you can get down to two choices, it is often worthwhile to guess from those. Once you've chosen an answer, move on and don't dwell on any that you skipped or had to hurry through. If a question was taking too long, chances are it was one of the harder ones, so you weren't as likely to get it right anyway.

On the other hand, if you find yourself getting ahead of schedule, it may be beneficial to slow down a little. The more quickly you work, the more likely you are to make a careless mistake that will affect your score. You've budgeted time for each question, so don't be afraid to spend that time. Practice an efficient but careful pace to get the most out of the time you have.

Secret Key #5 – Have a Plan for Guessing

When you're taking the test, you may find yourself stuck on a question. Some of the answer choices seem better than others, but you don't see the one answer choice that is obviously correct. What do you do?

The scenario described above is very common, yet most test takers have not effectively prepared for it. Developing and practicing a plan for guessing may be one of the single most effective uses of your time as you get ready for the exam.

In developing your plan for guessing, there are three questions to address:

- When should you start the guessing process?
- How should you narrow down the choices?
- Which answer should you choose?

When to Start the Guessing Process

Unless your plan for guessing is to select C every time (which, despite its merits, is not what we recommend), you need to leave yourself enough time to apply your answer elimination strategies. Since you have a limited amount of time for each question, that means that if you're going to give yourself the best shot at guessing correctly, you have to decide quickly whether or not you will guess.

Of course, the best-case scenario is that you don't have to guess at all, so first, see if you can answer the question based on your knowledge of the subject and basic reasoning skills. Focus on the key words in the question and try to jog your memory of related topics. Give yourself a chance to bring the knowledge to mind, but once you realize that you don't have (or you can't access) the knowledge you need to answer the question, it's time to start the guessing process.

It's almost always better to start the guessing process too early than too late. It only takes a few seconds to remember something and answer the question from knowledge. Carefully eliminating wrong answer choices takes longer. Plus, going through the process of eliminating answer choices can actually help jog your memory.

Summary: Start the guessing process as soon as you decide that you can't answer the question based on your knowledge.

How to Narrow Down the Choices

The next chapter in this book (**Test-Taking Strategies**) includes a wide range of strategies for how to approach questions and how to look for answer choices to eliminate. You will definitely want to read those carefully, practice them, and figure out which ones work best for you. Here though, we're going to address a mindset rather than a particular strategy.

Your odds of guessing an answer correctly depend on how many options you are choosing from.

Number of options left	5	4	3	2	1
Odds of guessing correctly	20%	25%	33%	50%	100%

You can see from this chart just how valuable it is to be able to eliminate incorrect answers and make an educated guess, but there are two things that many test takers do that cause them to miss out on the benefits of guessing:

- Accidentally eliminating the correct answer
- Selecting an answer based on an impression

We'll look at the first one here, and the second one in the next section.

To avoid accidentally eliminating the correct answer, we recommend a thought exercise called **the $5 challenge**. In this challenge, you only eliminate an answer choice from contention if you are willing to bet $5 on it being wrong. Why $5? Five dollars is a small but not insignificant amount of money. It's an amount you could afford to lose but wouldn't want to throw away. And while losing $5 once might not hurt too much, doing it twenty times will set you back $100. In the same way, each small decision you make—eliminating a choice here, guessing on a question there—won't by itself impact your score very much, but when you put them all together, they can make a big difference. By holding each answer choice elimination decision to a higher standard, you can reduce the risk of accidentally eliminating the correct answer.

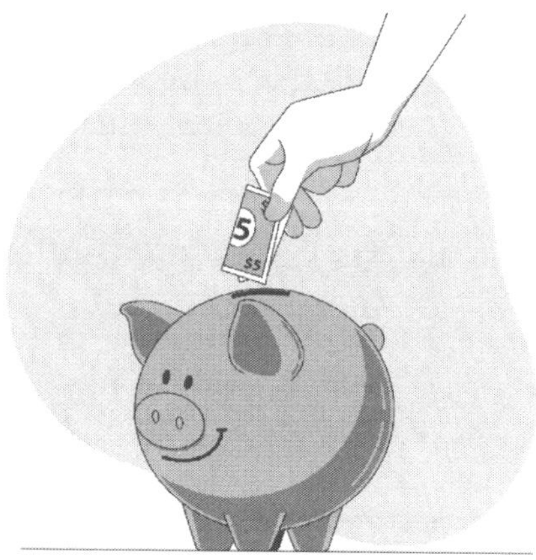

The $5 challenge can also be applied in a positive sense: If you are willing to bet $5 that an answer choice *is* correct, go ahead and mark it as correct.

Summary: Only eliminate an answer choice if you are willing to bet $5 that it is wrong.

Which Answer to Choose

You're taking the test. You've run into a hard question and decided you'll have to guess. You've eliminated all the answer choices you're willing to bet $5 on. Now you have to pick an answer. Why do we even need to talk about this? Why can't you just pick whichever one you feel like when the time comes?

The answer to these questions is that if you don't come into the test with a plan, you'll rely on your impression to select an answer choice, and if you do that, you risk falling into a trap. The test writers know that everyone who takes their test will be guessing on some of the questions, so they intentionally write wrong answer choices to seem plausible. You still have to pick an answer though, and if the wrong answer choices are designed to look right, how can you ever be sure that you're not falling for their trap? The best solution we've found to this dilemma is to take the decision out of your hands entirely. Here is the process we recommend:

Once you've eliminated any choices that you are confident (willing to bet $5) are wrong, select the first remaining choice as your answer.

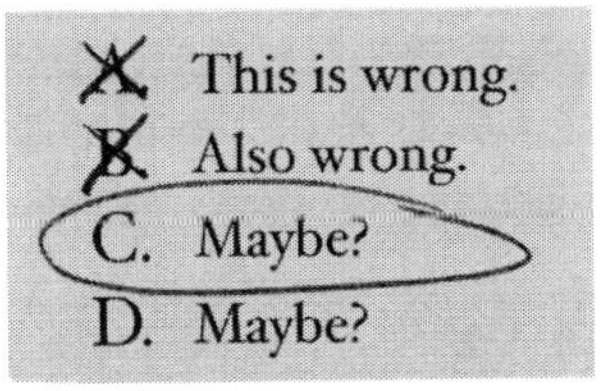

Whether you choose to select the first remaining choice, the second, or the last, the important thing is that you use some preselected standard. Using this approach guarantees that you will not be enticed into selecting an answer choice that looks right, because you are not basing your decision on how the answer choices look.

This is not meant to make you question your knowledge. Instead, it is to help you recognize the difference between your knowledge and your impressions. There's a huge difference between thinking an answer is right because of what you know, and thinking an answer is right because it looks or sounds like it should be right.

Summary: To ensure that your selection is appropriately random, make a predetermined selection from among all answer choices you have not eliminated.

Test-Taking Strategies

This section contains a list of test-taking strategies that you may find helpful as you work through the test. By taking what you know and applying logical thought, you can maximize your chances of answering any question correctly!

It is very important to realize that every question is different and every person is different: no single strategy will work on every question, and no single strategy will work for every person. That's why we've included all of them here, so you can try them out and determine which ones work best for different types of questions and which ones work best for you.

Question Strategies

✓ Read Carefully

Read the question and the answer choices carefully. Don't miss the question because you misread the terms. You have plenty of time to read each question thoroughly and make sure you understand what is being asked. Yet a happy medium must be attained, so don't waste too much time. You must read carefully and efficiently.

✓ Contextual Clues

Look for contextual clues. If the question includes a word you are not familiar with, look at the immediate context for some indication of what the word might mean. Contextual clues can often give you all the information you need to decipher the meaning of an unfamiliar word. Even if you can't determine the meaning, you may be able to narrow down the possibilities enough to make a solid guess at the answer to the question.

✓ Prefixes

If you're having trouble with a word in the question or answer choices, try dissecting it. Take advantage of every clue that the word might include. Prefixes can be a huge help. Usually, they allow you to determine a basic meaning. *Pre-* means before, *post-* means after, *pro-* is positive, *de-* is negative. From prefixes, you can get an idea of the general meaning of the word and try to put it into context.

✓ Hedge Words

Watch out for critical hedge words, such as *likely, may, can, often, almost, mostly, usually, generally, rarely,* and *sometimes*. Question writers insert these hedge phrases to cover every possibility. Often an answer choice will be wrong simply because it leaves no room for exception. Be on guard for answer choices that have definitive words such as *exactly* and *always*.

✓ Switchback Words

Stay alert for *switchbacks*. These are the words and phrases frequently used to alert you to shifts in thought. The most common switchback words are *but, although,* and *however*. Others include *nevertheless, on the other hand, even though, while, in spite of, despite,* and *regardless of*. Switchback words are important to catch because they can change the direction of the question or an answer choice.

✓ Face Value

When in doubt, use common sense. Accept the situation in the problem at face value. Don't read too much into it. These problems will not require you to make wild assumptions. If you have to go beyond creativity and warp time or space in order to have an answer choice fit the question, then you should move on and consider the other answer choices. These are normal problems rooted in reality. The applicable relationship or explanation may not be readily apparent, but it is there for you to figure out. Use your common sense to interpret anything that isn't clear.

Answer Choice Strategies

✓ Answer Selection

The most thorough way to pick an answer choice is to identify and eliminate wrong answers until only one is left, then confirm it is the correct answer. Sometimes an answer choice may immediately seem right, but be careful. The test writers will usually put more than one reasonable answer choice on each question, so take a second to read all of them and make sure that the other choices are not equally obvious. As long as you have time left, it is better to read every answer choice than to pick the first one that looks right without checking the others.

✓ Answer Choice Families

An answer choice family consists of two (in rare cases, three) answer choices that are very similar in construction and cannot all be true at the same time. If you see two answer choices that are direct opposites or parallels, one of them is usually the correct answer. For instance, if one answer choice says that quantity *x* increases and another either says that quantity *x* decreases (opposite) or says that quantity *y* increases (parallel), then those answer choices would fall into the same family. An answer choice that doesn't match the construction of the answer choice family is more likely to be incorrect. Most questions will not have answer choice families, but when they do appear, you should be prepared to recognize them.

✓ Eliminate Answers

Eliminate answer choices as soon as you realize they are wrong, but make sure you consider all possibilities. If you are eliminating answer choices and realize that the last one you are left with is also wrong, don't panic. Start over and consider each choice again. There may be something you missed the first time that you will realize on the second pass.

✓ Avoid Fact Traps

Don't be distracted by an answer choice that is factually true but doesn't answer the question. You are looking for the choice that answers the question. Stay focused on what the question is asking for so you don't accidentally pick an answer that is true but incorrect. Always go back to the question and make sure the answer choice you've selected actually answers the question and is not merely a true statement.

✓ Extreme Statements

In general, you should avoid answers that put forth extreme actions as standard practice or proclaim controversial ideas as established fact. An answer choice that states the "process should be used in certain situations, if..." is much more likely to be correct than one that states the "process should be discontinued completely." The first is a calm rational statement and doesn't even make a definitive, uncompromising stance, using a hedge word *if* to provide wiggle room, whereas the second choice is far more extreme.

⊘ Benchmark

As you read through the answer choices and you come across one that seems to answer the question well, mentally select that answer choice. This is not your final answer, but it's the one that will help you evaluate the other answer choices. The one that you selected is your benchmark or standard for judging each of the other answer choices. Every other answer choice must be compared to your benchmark. That choice is correct until proven otherwise by another answer choice beating it. If you find a better answer, then that one becomes your new benchmark. Once you've decided that no other choice answers the question as well as your benchmark, you have your final answer.

⊘ Predict the Answer

Before you even start looking at the answer choices, it is often best to try to predict the answer. When you come up with the answer on your own, it is easier to avoid distractions and traps because you will know exactly what to look for. The right answer choice is unlikely to be word-for-word what you came up with, but it should be a close match. Even if you are confident that you have the right answer, you should still take the time to read each option before moving on.

General Strategies

⊘ Tough Questions

If you are stumped on a problem or it appears too hard or too difficult, don't waste time. Move on! Remember though, if you can quickly check for obviously incorrect answer choices, your chances of guessing correctly are greatly improved. Before you completely give up, at least try to knock out a couple of possible answers. Eliminate what you can and then guess at the remaining answer choices before moving on.

⊘ Check Your Work

Since you will probably not know every term listed and the answer to every question, it is important that you get credit for the ones that you do know. Don't miss any questions through careless mistakes. If at all possible, try to take a second to look back over your answer selection and make sure you've selected the correct answer choice and haven't made a costly careless mistake (such as marking an answer choice that you didn't mean to mark). This quick double check should more than pay for itself in caught mistakes for the time it costs.

⊘ Pace Yourself

It's easy to be overwhelmed when you're looking at a page full of questions; your mind is confused and full of random thoughts, and the clock is ticking down faster than you would like. Calm down and maintain the pace that you have set for yourself. Especially as you get down to the last few minutes of the test, don't let the small numbers on the clock make you panic. As long as you are on track by monitoring your pace, you are guaranteed to have time for each question.

⊘ Don't Rush

It is very easy to make errors when you are in a hurry. Maintaining a fast pace in answering questions is pointless if it makes you miss questions that you would have gotten right otherwise. Test writers like to include distracting information and wrong answers that seem right. Taking a little extra time to avoid careless mistakes can make all the difference in your test score. Find a pace that allows you to be confident in the answers that you select.

⊘ KEEP MOVING

Panicking will not help you pass the test, so do your best to stay calm and keep moving. Taking deep breaths and going through the answer elimination steps you practiced can help to break through a stress barrier and keep your pace.

Final Notes

The combination of a solid foundation of content knowledge and the confidence that comes from practicing your plan for applying that knowledge is the key to maximizing your performance on test day. As your foundation of content knowledge is built up and strengthened, you'll find that the strategies included in this chapter become more and more effective in helping you quickly sift through the distractions and traps of the test to isolate the correct answer.

Now that you're preparing to move forward into the test content chapters of this book, be sure to keep your goal in mind. As you read, think about how you will be able to apply this information on the test. If you've already seen sample questions for the test and you have an idea of the question format and style, try to come up with questions of your own that you can answer based on what you're reading. This will give you valuable practice applying your knowledge in the same ways you can expect to on test day.

Good luck and good studying!

Scientific and Engineering Practices

Safety

SAFETY PROCEDURES

Everyone working in a lab setting should be careful to follow these rules to protect themselves and others from injury or accidents.

- Students should wear a **lab apron** and **safety goggles**.
- **Loose** or **dangling** clothing and jewelry, necklaces, and earrings should not be worn.
- **Long hair** should be tied back.
- Care should always be taken not to **splash chemicals**.
- **Open-toed shoes** such as sandals and flip-flops should not be worn, nor should wrist watches.
- **Glasses** are preferable to contact lenses since the latter carries a risk of chemicals getting caught between the lens and the eye.
- Students should always be **supervised** during an experiment.
- The area where the experiment is taking place and the surrounding floor should be **free of clutter**.
- **Food** and **drink** should also not be allowed in a lab setting.
- **Cords** should not be allowed to **dangle** from work stations.
- There should be no **rough-housing** in the lab.
- **Wash hands** before and after performing laboratory activities.

SAFETY GLOVES

There are many types of **gloves** available to help protect the skin from cuts, burns, and chemical splashes. There are many factors to consider when choosing a glove. For example, gloves that are highly protective may limit grip or accuracy. Some gloves may not offer appropriate protection against a specific chemical. Disposable latex, vinyl, or nitrile gloves are usually appropriate for most circumstances, and offer protection from incidental splashes and contact.

LABORATORY ACCIDENTS

Accidents happen in labs, so it is important to know how to clean up, stay safe, and report the accident to the teacher. Any spills or accidents should be **reported** to the teacher so that the teacher can determine the safest clean-up method. The student should start to wash off a **chemical** spilled on the skin while reporting the incident. Some spills may require removal of contaminated clothing and use of a **safety shower**. Broken glass should be disposed of in a designated container. If a piece of clothing catches fire, the individual wearing the clothing should walk to the safety shower and use it to extinguish the flames. A fire blanket may be used to smother a **lab fire**. A fire extinguisher, phone, spill neutralizers, and a first aid box are other types of **safety equipment** found in the lab. Students should be familiar with **routes** out of the room and the building in case of fire.

Science Knowledge and Scientific Process

Scientific knowledge is knowledge about the world that we understand by observing and testing. The **scientific process** is how we try to gain scientific knowledge. The steps are to make an initial observation, create a hypothesis, test the hypothesis with an experiment, draw conclusions from the experiment, then start over with new observations.

OBSERVATIONS

An **observation** is something specific you notice about the world or an event. For instance, a very obvious event that takes place is if we leave milk out on the counter overnight, the milk will go bad. We observe this event by using our five senses: sight, taste, smell, touch, and sound. When milk goes bad, there are several ways it changes, including the way it looks and smells.

ASKING QUESTIONS

Observations may inspire new questions. These questions can be used to form a hypothesis, which can be used to design an experiment. When trying to gain scientific knowledge, it is best to ask questions that are clear and specific, or well defined. This can make answering the question easier. When a question is not clear or is not specific, it may have several possible answers. It also may have an answer that only leads to more questions. For example, the question "what will happen if you drop an egg" is not a well-defined question. The result of dropping an egg depends on many factors, and there are many possible answers to the question. However, the question "what will happen if you drop an uncooked egg on concrete" is a well-defined question. The question is specific and clear, and it can be tested using a simple experiment.

HYPOTHESIS

After making observations, the next step is to make an educated guess about what might happen in a certain event. This is also known as a **hypothesis**. A hypothesis about milk might be that if I keep milk above a certain temperature for a long time, it will go bad. Hypotheses need to be specific and testable so it can be determined whether they are true or false.

EXPERIMENT

The step after creating a hypothesis is to check if it is true by testing it with an **experiment**. An experiment is a type of investigation that follows a list of steps in order to find evidence that supports or does not support a hypothesis. Experiments need to include a controlled, specific situation that reveals whether or not the hypothesis is correct. Experiments need to ask questions about **facts**, rather than **opinions**. The hypothesis can help the experimenter decide what steps should be included in the experiment.

VARIABLES

Experiments include variables. **Variables** are parts of an experiment that may be changed or kept the same during the experiment, and some variables affect others. An **independent variable** is a variable that the experimenter will change during the experiment, and a **dependent variable** is a variable that changes in response to changes to an independent variable. Independent variables are changed, and dependent variables are observed. Variables need to be controlled as much as possible to get the most specific results.

EXPERIMENTS TESTING ONE VARIABLE

Experiments should be designed to only test the effects of one variable. This helps experimenters make sure that the cause of the experiment's results is clear and certain. To make sure that the experiment only tests one variable, the experiment should only have one independent variable. For example, the height that sunflowers would grow in different circumstances depends on many factors that could be changed by the experimenter. Sunflowers are affected by the amount of water and sunlight they receive, the soil in which they are planted, the amount of space they have to grow, and the temperature of the air around them. If more than one of these factors are independent variables, then the experimenter may not be able to tell which factor affected growth of the sunflower. If only one of these factors is an independent variable, then the experimenter will know which factor affected the sunflower, since it will be the only thing that was changed by the experimenter.

CONSTANTS AND CONTROLS

To ensure that there is only one independent variable, other factors cannot be changed by the experimenter. These factors are called **constants**. Another aspect of an experiment is the control. A **control** is an example where the independent variable is not manipulated. The group in which the independent variable was manipulated can then be compared to the control group.

COLLECTING INFORMATION DURING AN EXPERIMENT

An experimenter makes observations during an experiment, writes them down, and then uses those observations to **draw conclusions**. Observations made during an experiment become **data**. Accurately **measuring** and **recording** data is necessary for an effective scientific investigation. **Accuracy** describes how close a measurement is to the true value. Not accurately measuring during an experiment may cause an experimenter to misunderstand the results of the experiment. For example, investigating how much plants grow when given different amounts of water would require the experimenter to accurately measure and record how much water each plant received and how much each plant grew daily. When an experimenter is writing and sharing the conclusion of a scientific investigation, accurate information and specific observations are needed so the experiment can be repeated and considered to be reliable.

Review Video: What is the Scientific Method?
Visit mometrix.com/academy and enter code: 191386

DIRECT AND INDIRECT EVIDENCE

Evidence refers to the data and facts that **support** a hypothesis or explanation. There are different types of evidence that are used when analyzing the result of a scientific investigation:

- **Direct evidence** is any evidence that clearly supports the hypothesis or explanation and does not need any additional support. A fossil, which is the preserved remains of a past organism, is an example of direct evidence. Fossils show that certain organisms existed long ago.
- **Indirect evidence** is evidence that is used with additional evidence and inferences to support the hypothesis or explanation. For example, identifying why an animal learned a new behavior would require looking at indirect evidence. One might look at the animal's original behavior, its new behavior, and any events that occurred before the animal began to behave differently. These pieces of evidence could be used to infer a reason for the animal's adaptation, but this evidence does not clearly prove why the animal learned the new behavior.

Both types of evidence are important for supporting a hypothesis. Direct evidence can be used to prove exactly what happened, while indirect evidence can be used to make connections that will lead to a better understanding and explanation of what took place during the investigation.

Repeating Experiments

Experiments should be **repeated** to see if the same result happens every time. For example, if you can get the same result in only three out of ten attempts, then the result is not reliable, and the experiment should be changed and tried again.

The results of an investigation that is only run once are not completely reliable. When an investigation is run the exact same way again and again and the result is the same every time, then the result is more likely to be considered reliable. If the investigation has only been completed once, or even just a few times, the results may be a result of chance or an error. When an investigation can be repeated, there is an opportunity for other discoveries to occur. An investigation may be repeated and produce different results. While those results may not be reliable, they can still lead to a new discovery or idea that can be investigated. The results of a repeated investigation may become a scientific theory, which is an explanation of an event that has become widely accepted by the science community.

Reflect

After an experiment is complete, the scientist needs to **reflect** on what happened by asking questions:

- Did the experiment answer the question?
- Was the hypothesis true or false?
- Did repeated attempts produce the same results?
- Is there anything I should change for future attempts?
- What did I learn?
- What questions can I ask now that I have learned something?

Repeat the Process

After the experiment and reflection is over, the scientist hopefully learned something and can ask new questions. These questions should be more specific, or the experiment should be improved to get better answers.

Sharing Results

The result of a scientific investigation is valuable information that others may want to know. These results may help other scientists learn new information that is related to their own research and investigations. They may also help others if the information changes the way people may look at an everyday product or activity. However, results can be communicated in different ways. When the results of an experiment can be expressed using numbers, different graphs and charts can share the information in a way that emphasizes patterns and the investigator's conclusions. Lab reports are also useful when sharing the results of an investigation. A lab report includes a description of the investigation, so the reader will know which methods were used in the investigation and the results of the investigation.

Lab Reports

A **lab report** should begin with a statement describing the problem or question being investigated. It should also include the hypothesis and list the variables in the experiment. Both a list of all materials used and the detailed procedure that was followed during the investigation are important to include. These will allow other scientists to replicate the investigation. A lab report should also list the results of an investigation. Any data tables or observations that are part of the results need to be labeled correctly and include accurate measurements and calculations. Finally, a lab report should conclude with an explanation that uses the evidence that was found during the investigation to state whether the hypothesis was supported or not supported. Also, the conclusion should include a detailed summary of the data collected during the investigation, and it should describe any errors that might have occurred during the investigation.

VENN DIAGRAMS

One helpful way of sorting out information when comparing two things is a **Venn diagram.** This tool is made by drawing two circles that overlap. It is used to display which qualities of one thing are similar or different from the qualities of another thing. In the following diagram, notice how fish and frogs are similar in some ways, but different in others.

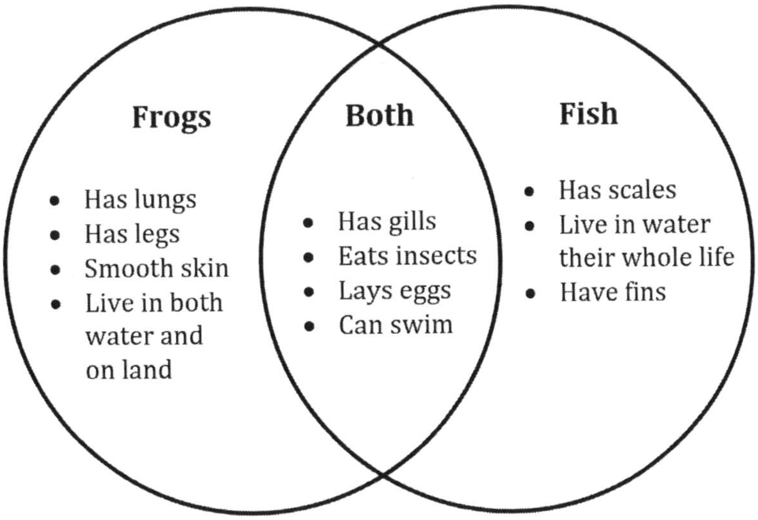

FREQUENCY TABLES

Frequency tables show how frequently each unique value appears in a set. A **relative frequency table** is one that shows the proportions of each unique value compared to the entire set. Relative frequencies are given as percentages; however, the total percent for a relative frequency table will not necessarily equal 100 percent due to rounding. An example of a frequency table with relative frequencies is below.

Favorite Color	Frequency	Relative Frequency
Blue	4	13%
Red	7	22%
Green	3	9%
Purple	6	19%
Cyan	12	38%

CIRCLE CHARTS

Circle charts, also known as **pie charts**, show what portion of a whole set of data is made of a specific type of data. The circle chart is divided into sections by drawing lines to create wedges. The percentage of the circle that each wedge takes up is equal to the percentage of the whole data set that is taken up by that type of data. When complete, a circle chart often looks like a pie cut into uneven wedges. That is why it may also be called a pie chart.

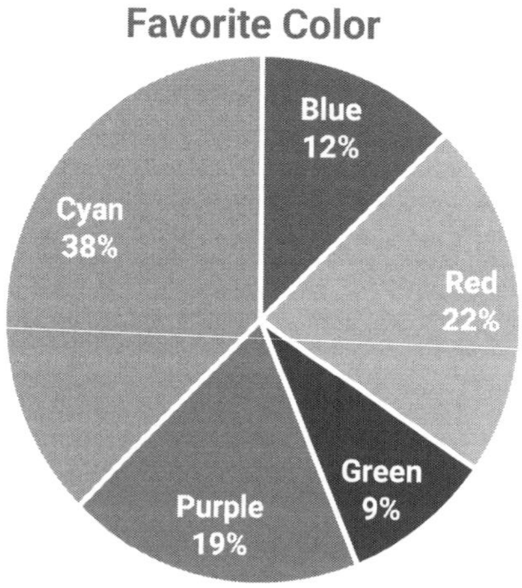

PICTOGRAPHS

A **pictograph** is a graph that uses pictures or symbols to represent data. Each pictograph must have a key that defines the picture or symbol and tells what quantity each picture or symbol represents. Pictures or symbols on a pictograph are not always shown as whole pictures or symbols. In this case, the fraction of the picture or symbol shown represents the same fraction of the quantity the whole picture or symbol stands for. For example, a row with $3\frac{1}{2}$ ears of corn, where each ear of corn represents 100 stalks of corn in a field, would equal $3\frac{1}{2} \times 100 = 350$ stalks of corn in the field.

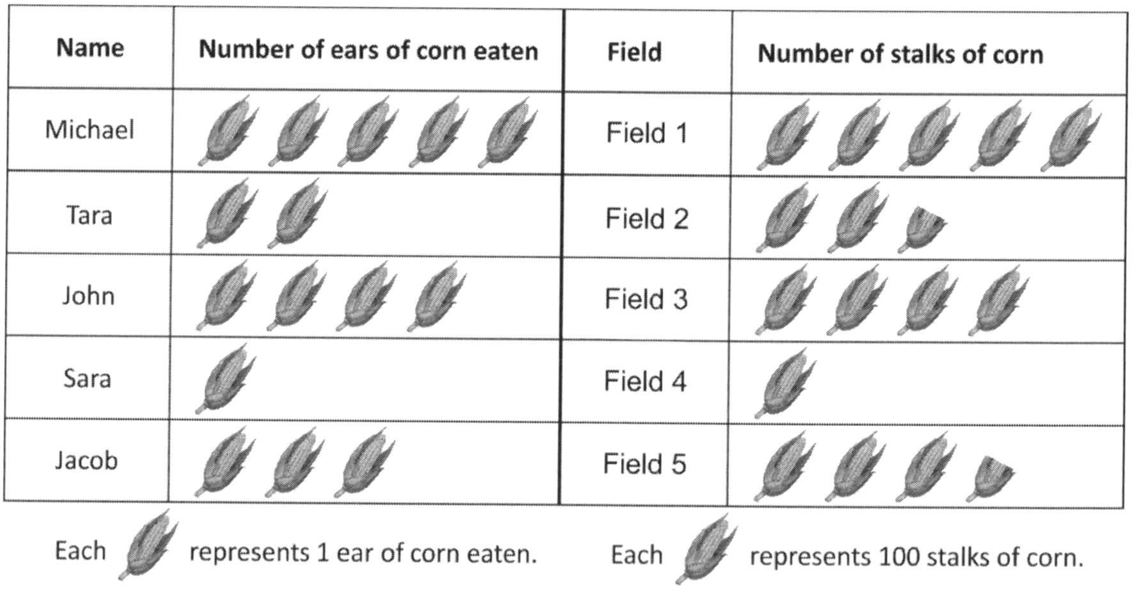

Review Video: Pictographs
Visit mometrix.com/academy and enter code: 147860

LINE GRAPHS

Line graphs have one or more lines of varying styles (solid or broken) to show the different values for a set of data. The individual data are represented as ordered pairs, much like on a coordinate plane. In this case, the *x*- and *y*-axes represent units, such as dollars or time. The individual plotted points are joined by line segments to show whether the value of the data is increasing (line sloping upward), decreasing (line sloping downward), or staying the same (horizontal line). Multiple sets of data can be graphed on the same line graph so they can be easily compared. An example of this would be graphing test scores for different students over the same time period. The graph will show which students had the greatest increase or decrease in their scores from year to year.

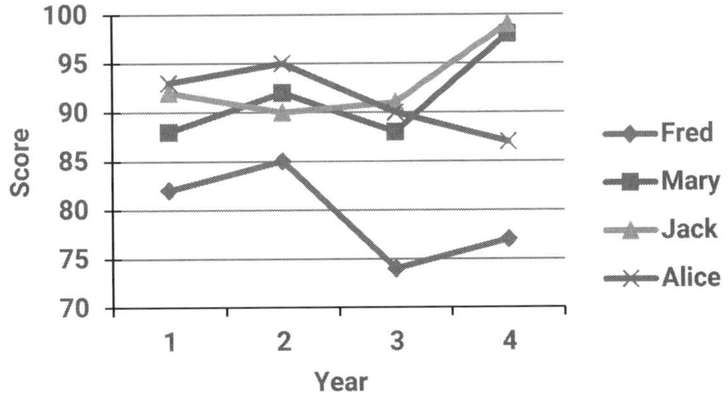

Review Video: How to Create a Line Graph
Visit mometrix.com/academy and enter code: 480147

LINE PLOTS

A **line plot**, also known as a **dot plot**, has plotted points that are not connected by line segments. In this type of graph, the horizontal axis lists the different possible values for the data, and the vertical axis lists the number of times each value occurs. A single dot is graphed for each value to show the number of times it occurs. This graph is more closely related to a bar graph than to a line graph. Do not connect the dots in a line plot, or it will misrepresent the data.

Review Video: Line Plot
Visit mometrix.com/academy and enter code: 754610

STEM AND LEAF PLOTS

A **stem and leaf plot** is useful for depicting groups of data that fall into a range of values. Each piece of data is separated into two parts: the first, or left, part is called the stem; the second, or right, part is called the leaf. Each stem is listed in a column from smallest to largest. Each leaf that has the common stem is listed in that stem's row from smallest to largest. For example, in a set of two-digit numbers, the digit in the tens place is the stem, and the digit in the ones place is the leaf. With a stem and leaf plot, you can easily see which subset of numbers (10s, 20s, 30s, etc.) is the largest. This information is also readily available by looking at a histogram, but a stem and leaf plot also allows you to look closer and see exactly which values fall in that range. Using a sample set of test scores (82, 88, 92, 93, 85, 90, 92, 95, 74, 88, 90, 91, 78, 87, 98, 99), we can assemble a stem and leaf plot like the one below.

Test Scores

7	4	8							
8	2	5	7	8	8				
9	0	0	1	2	2	3	5	8	9

Review Video: Stem-and-Leaf Plots
Visit mometrix.com/academy and enter code: 302339

Bar Graphs

A **bar graph** is similar to a line plot in the way the data is organized on the graph. Bar graphs can be vertical or horizontal, but both axes must have their categories labelled for the graph to be useful. Rather than placing a single dot to mark the point of the data's value, a bar, or thick line, is drawn from zero to the exact value of the data. The value may be a number, percentage, or other numerical value. Longer bar lengths correspond to greater data values. To read a bar graph, read the labels for the axes to find the units being represented. Then, look where the bars end in relation to the scale given on the other axis and find the value of the data.

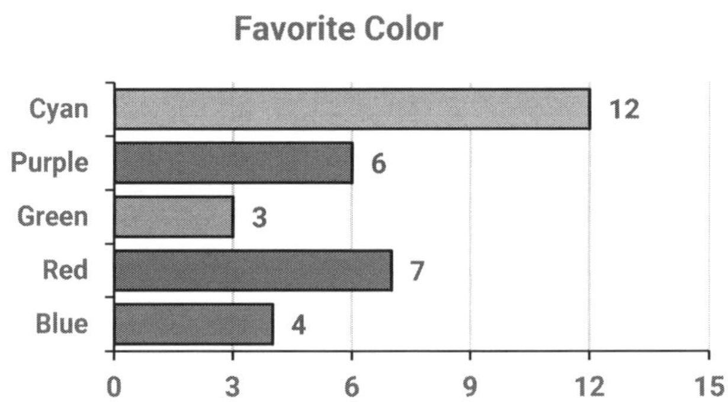

Evaluating Scientific Explanations

The results of an investigation and the conclusions that scientists make after studying data are used to create scientific explanations. **Explanations** seek to describe why natural events happen and provide new details about scientific discoveries. However, some explanations are not reliable. When new scientific explanations are made, it is important to evaluate them to see if they are reliable and possible. Important factors to think about and questions to ask when evaluating a scientific explanation are:

- **Logic:** Does the explanation make sense? Is the explanation reasonable? Does it seem possible?
- **Question or Problem:** Was the question or problem that the investigation studied well-defined? Did the investigation results answer that question?
- **Experimental methods:** Is the method used in the experiment reliable? Did the experimenter record their process? Was the experiment performed properly and carefully?
- **Tools:** Did the tools used take accurate measurements? Were the correct tools used?
- **Evidence:** Is the explanation based on direct or indirect evidence? Does the evidence support the explanation well?
- **Research:** Did the investigator use reliable, scientific resources when doing research related to the investigation or experiment?

Scientific Models

Scientists often create **models** to help express and understand scientific ideas. A model is a representation of an idea that is used to experience something that cannot be experienced directly. For instance, we cannot see and experience evaporation directly, but we can easily draw out the process of evaporation, condensation, and precipitation. Having a drawing helps us to understand the concept since we cannot experience it directly. Models can be three-dimensional, such as something you can hold, or they can be simple diagrams showing a process. For instance, a planetarium is a large model that helps someone see the size of planets and stars and how they interact. More typical models in the classroom include baking soda and vinegar volcanoes, solar system mobiles, or molecules made from foam and straws. Other models can include a diagram showing the water cycle, round objects showing the movement and location of the Sun, Moon, and Earth, or layers of different materials showing how sedimentary rocks and fossil fuels are made.

Limitations of Models

Models can be helpful for experiencing a scientific process that normally could not be observed directly, but they are limited. Models are usually used to display a very large thing in a small way, such as showing a volcano or the Earth in a way that a human can look at it and touch it. Models are also often used to express very small things or invisible things in a more observable way. Changing the size or the materials used to express something limits how similar a model can be to the real thing. For instance, if a model of the water cycle used cotton balls to show clouds, it limits our understanding of clouds. Clouds are not made of dry, solid cotton, but are made up of vapor filled with water. Volcanos made from clay that use baking soda and vinegar do not demonstrate the extreme heat or mass of volcanic materials. Similarly, a mobile of the solar system cannot express how extremely large the Sun and planets are. When using a model to understand a concept, the observer needs to be aware that they are probably only getting a clear picture of a small part of the concept, not gaining full understanding from the model.

Measurement

Tools for Measuring and Observation

Purpose	Tool
Measuring length	Rulers, meter sticks
Measuring weight or mass	Spring scales, pan balances
Measuring volume	Beakers, graduated cylinders
Measuring time	Timers, clocks
Measuring temperature	Thermometer
Recording information	Journals, notebooks
Observing animals and plants	Terrariums, aquariums, collecting nets
Observing weather	Rain gauges, wind vanes
Observing light	Prisms, mirrors
Observing magnetism	Magnets
Observing effects of heat	Hot plates
Models for understanding concepts	Sun-Earth-Moon system models, volcano models, water cycle models
Visual observation	cameras, hand lenses, microscopes
Making calculations	Calculators, computers

Metric System

The **metric system** is the accepted standard of measurement in the scientific community. **Standardization** is important because it allows the results of experiments to be compared and reproduced without the need to convert measurements. The metric system uses similar conversions between all types of measurement, including length, mass, volume, time, and temperature.

Customary and Metric (Scientific) Units of Measurement

The customary system commonly used in the United States is not based on consistent smaller units. Thus, 12 inches equal 1 foot, 3 feet equal 1 yard, and 5,280 feet equal 1 mile. The metric system used in science and in most countries is based on units of 10. Therefore, 1000 millimeters and 100 centimeters equal 1 meter, and 1,000 meters equal 1 kilometer. Most units of measurement follow the pattern of being based on units of 10, with a notable exception being temperature. The following table shows some different units.

Unit	Customary System	Metric System
Length	Inch, foot, mile	Centimeter, meter, kilometer
Mass, weight	Net weight ounce, pound	Gram, kilogram, newton
Volume	Fluid ounce, pint, quart	Milliliter, liter
Temperature	Fahrenheit degree	Celsius degree

Basic Units of Measurement

The **metric system** is generally accepted as the preferred measurement system. Having a **universal standard** allows individuals to interpret measurements more easily, regardless of where they are located. The basic units of measurement are the **meter**, which measures length; the **liter**, which measures volume; and the **gram**, which measures mass. The metric system starts with a base unit and increases or decreases in units of 10. The prefix and the base unit combined are used to indicate an amount. For example, deka- is 10 times the base unit. A decameter, or dekameter, is 10 meters; a decaliter, or dekaliter, is 10 liters; and a decagram, or dekagram, is 10 grams. The prefix hecto- refers to 100 times the base amount and kilo- is 1,000 times the base amount. The prefixes that indicate a fraction of the base unit are deci-, which is 1/10 of the base unit; centi-, which is 1/100 of the base unit; and milli-, which is 1/1000 of the base unit.

Review Video: Metric System Conversions
Visit mometrix.com/academy and enter code: 163709

Common Prefixes

The prefixes for **multiples** are as follows:

Deca or Deka	(da)	10
Hecto	(h)	100
Kilo	(k)	1,000
Mega	(M)	100,000

The prefixes for **subdivisions** are as follows:

Deci	(d)	1/10
Centi	(c)	1/100
Milli	(m)	1/1,000
Micro	(μ)	1/100,000

Fahrenheit and Celsius Temperature Scales

In the Fahrenheit scale the point where water freezes and ice melts is set at 32°, and the point where water boils and water vapor condenses is 212°. That means there is a difference of 180° between the freezing and boiling points of water. In the Celsius scale, the freezing/melting point is set at 0°and the boiling/condensation point at 100°, making this scale much easier to use.

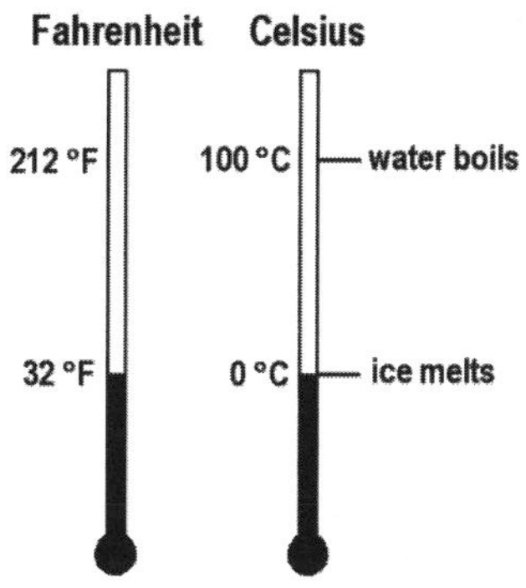

Contributions of Scientists

- **Sir Francis Bacon** – contributed many ideas that helped develop the modern scientific method.
- **Galileo Galilei**– one of the first people to use a telescope and contributed ideas about how gravity works. He also contributed the idea that the Earth revolved around the Sun, and not the other way around.
- **Leonardo da Vinci** – known for his contributions to art and to science, he was an avid inventor, designing a submarine, an armored tank, and several aircraft far before any of which were actually built.
- **Sir Isaac Newton** – best known for his contributions to physics. He described the principles of inertia and friction.
- **Louis Pasteur** – discovered bacteria, invented the process of pasteurization, and developed some of the first vaccines.
- **Thomas Edison** – invented many distinct devices, including the phonograph, improved the telephone, the lightbulb, and the kinetoscope, an early form of movie projector.
- **Albert Einstein** – studied theoretical physics with the use of mathematics and is particularly famous for his theory of relativity.

Force, Motion, Energy, and Matter

Energy

Performing work or moving an object is a function of **energy**. The **law of conservation of energy** states that energy cannot be created or destroyed. Energy can only be changed to another form. Therefore, energy is always part of a pattern, cycle, or system. Two common forms of energy are potential energy and kinetic energy. **Potential energy** is energy that is **stored** in matter. If an object or organism moves or is moved, its potential energy will become kinetic energy. **Kinetic energy** is the energy of **motion**. Therefore, energy used during movement is always kinetic energy. A change from one type of energy to another is called a **conversion**.

Mechanical Energy

Mechanical energy is the term for the total potential and kinetic energy of an object. Another way to say that is with the equation:

$$mechanical\ energy = potential\ energy + kinetic\ energy$$

Mechanical energy can be increased by increasing either the potential energy, the kinetic energy, or both. For example, lifting a block or squishing a spring increases potential energy. In contrast, throwing a ball or swinging a hammer increases kinetic energy. In each of those examples, mechanical energy is increased.

Thermal Energy

Thermal energy is energy that is produced when particles move and give off **heat**. The Sun is the main source of the thermal energy on Earth. Fire is another natural source of thermal energy. Many devices, such as hair dryers, can also produce thermal energy using parts that are designed to heat up when they are powered using electrical energy.

Electrical Energy

Electrical energy is energy that is produced when **electrons** or **electric charges** move from one place to another. Items that need **electricity**, such as lights and household appliances, use electrical energy.

Light and Sound Energy

Light travels in waves, and the light visible to humans is an example of **light energy**. Most of the light energy on Earth comes from sunlight or bulbs powered by electricity.

Sound energy is caused by **vibrations**. When something moves or has an electrical current moving through it, it will vibrate. This creates waves that move through the space around the object that is vibrating, and sound energy is generated as those waves move. The way the vibration sounds depends on characteristics of the wave. Machines such as alarm clocks or radios include parts that are designed to produce sound waves.

EXAMPLE

Which of these actions is an example of mechanical energy?

a. Warming your hands over a campfire
b. Using a lamp to read a book
c. Riding a bicycle
d. Listening to the radio

Choice A describes the transfer of heat from a fire to your hands. Heat is the movement of thermal energy, while mechanical energy is based on an object's motion or position. Choice A is not an example of mechanical energy.

Choice B involves a lamp, which converts electrical energy to produce light energy. Neither of these are examples of mechanical energy, so Choice B is not the answer.

Choice C involves someone applying a force to the pedals of a bicycle, causing the bicycle to move forward. Since mechanical energy has to do with the movement of an object, Choice C could be the answer.

Choice D describes a radio, which receives radio wave signals and converts them into sound energy. Sound energy does not involve the movement of an object, so Choice D is not the answer.

Choice C, riding a bicycle, is the only option that describes the energy of an object due to its motion or position, making it the correct answer.

Examples of Energy Conversions

The **law of conservation of energy** states that energy cannot be created nor destroyed; it can only be **converted,** or transformed from one form to another. In fact, many everyday actions require energy conversions. For example, batteries contain **chemical energy** in the form of charged materials. When you switch on a flashlight, it completes an electrical circuit, which converts the chemical energy from the battery into **electrical energy.** Then, the bulb on the flashlight converts this electrical energy into **light,** a form of **electromagnetic energy.** The human body also uses chemical energy by processing food to produce **mechanical energy** that can be used to move the muscles. Plugging in a space heater converts **electrical energy** into **thermal energy** or heat. Another example of energy conversion is a falling object. At its peak, an object has **gravitational potential energy**, or the potential to fall and gain speed. When an object is dropped, gravitational potential energy is converted into **kinetic energy**, the energy of motion. In some cases, it may seem like energy is being lost or destroyed, such as when a skateboarder slows down to a stop. However, in this case, the kinetic energy is instead being converted into thermal energy as a result of **friction**.

Motion and Force

Sir Isaac Newton observed three major laws that explain motion. The term **motion** refers to the movement of an object. The term **force** refers to physical energy that is applied to an object. Newton referred to objects that are not moving as being in a state of **rest**.

- **Newton's First Law of Motion** - a body will remain at **rest** or in **motion** until a **force** is applied to it.
    - A bowling ball continues to roll in the same direction it was thrown until friction eventually slows it down.
    - A ball sitting on the ground will not move unless a force is applied to it.
    - A wagon will remain still until it is pulled by a person or an animal.
- **Newton's Second Law of Motion** - the amount of **acceleration** is determined by the amount of **force** and the **mass** of the object being moved.
    - A soccer ball will accelerate more if it is kicked harder.

    - A shopping cart is much easier to push than a car because it has much less mass than a car does.

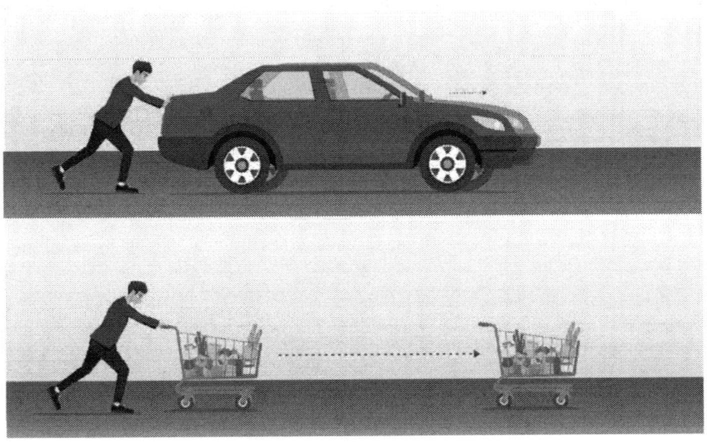

- **Newton's Third Law of Motion** – For every **action**, there is an equal and opposite **reaction**.
    - The floor applies force upward to hold up the weight of a person standing on the floor.
    - A dog pulls against a leash that its owner is holding.

Directions of Motion

Motion can be linear, rotational, or oscillating.

- **Linear motion** – motion in a single direction, such as a bowling ball rolling down the lane or a train moving in one direction down its tracks
- **Rotational motion** – rotating or spinning in a circular motion, such as a top spinning, a merry-go-round, or the Earth rotating around the Sun
- **Oscillating motion** – bouncing up and down repeatedly, such as a yo-yo on a string, the motion of a spring, or the rising and falling of waves

Mass and Strength of Force

An object with a large mass can be moved by strong forces, but may not be moved by a weak force. However, a weak force may move an object with little mass very easily. In general, a strong force will have a greater effect on an object than a weak force will.

GRAVITY

Gravity is a force that exists between all objects with matter. **Gravity** is a pulling force between objects, meaning that the forces on the objects point toward the opposite object. Gravity as we experience it is the force that the Earth exerts on objects, pulling them downward toward the center of the Earth. When Newton's third law is applied to gravity, the force pairs from gravity are shown to be equal in magnitude and opposite in direction.

Technically, all matter pulls other matter. The more massive the object, the more it pulls. The Earth is seen as our center of gravity because it is the most massive object nearby. Gravity is also reason that the Earth and other planets revolve around the Sun. The Sun is so massive that the Earth and all of the other bodies within the solar system are drawn to it and revolve around it as a result.

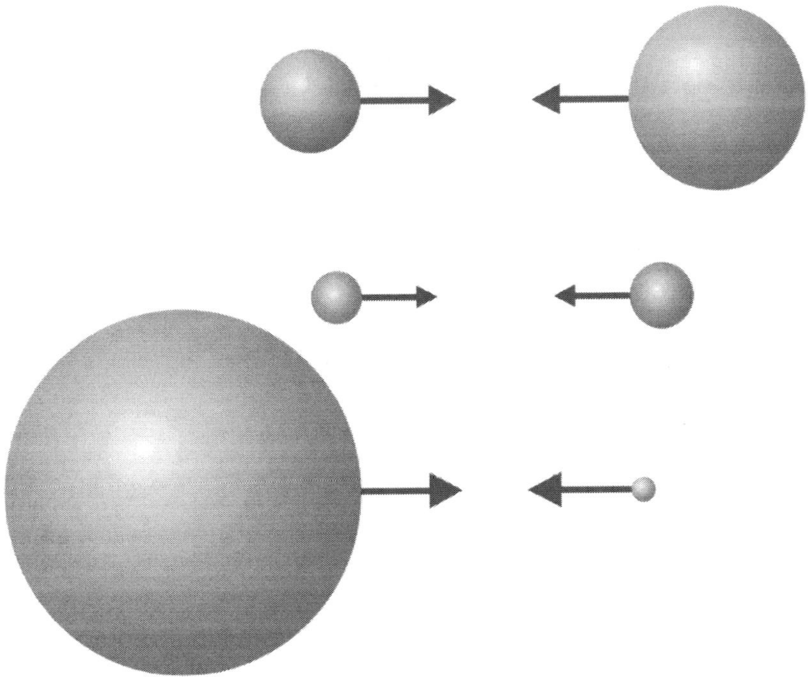

Friction

Friction is a resistance to motion between contacting surfaces. In order to illustrate the concept of friction, let us imagine a book resting on a table. As it sits, the force of its weight is equal to and opposite of the normal force. If, however, we were to exert a force on the book, attempting to push it to one side, a frictional force would arise, equal and opposite to our force. This kind of frictional force is known as static frictional force.

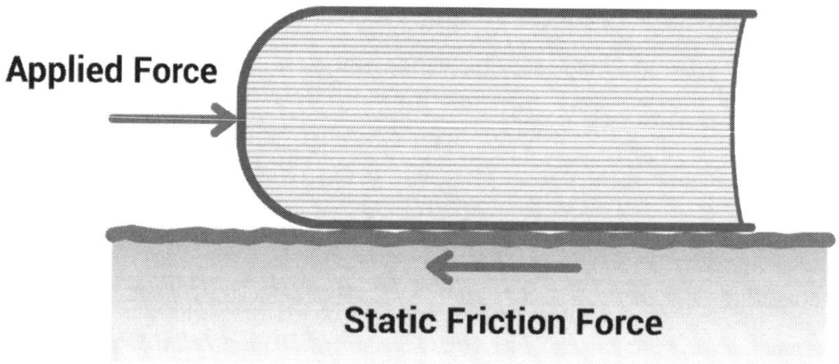

As we increase our force on the book, however, we will eventually cause it to accelerate in the direction of our force. At this point, the frictional force opposing us will be known as kinetic friction. For many combinations of surfaces, the magnitude of the kinetic frictional force is lower than that of the static frictional force, and consequently, the amount of force needed to maintain the movement of the book will be less than that needed to initiate the movement.

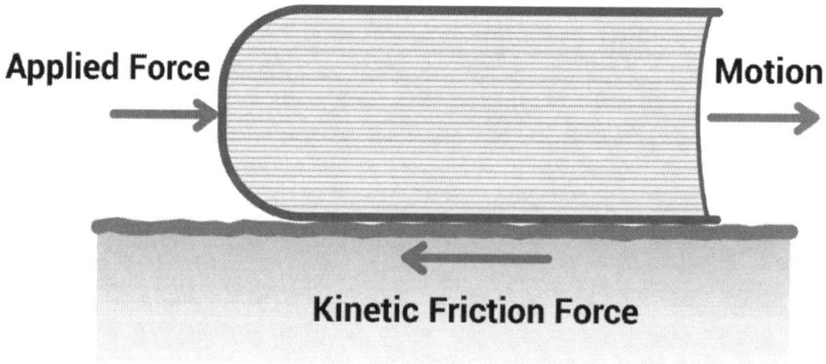

Review Video: Friction
Visit mometrix.com/academy and enter code: 716782

Intensive and Extensive Physical Properties of Matter

Intensive properties are properties that do not change when the size of a material changes. Density, melting point, freezing point, boiling point, color, chemical reactivity, luster, malleability, and electrical conductivity are examples of intensive properties. In contrast, **extensive properties** do change when the size of a material changes. Volume, mass, and weight are examples of extensive properties. Some intensive properties can be known just by looking at an object, such as color and state of matter. Other intensive and extensive properties can only be measured using special tools or by using tests to learn more about an object. Properties that must be measured or found using tests include density, mass, volume, solubility, magnetism, and conductivity.

States of Matter

The three states of matter are solid, liquid, and gas. A **solid** has an exact shape and an exact volume. In a **solid**, the molecules are close together and locked into place. A **liquid** has an exact volume but does not have an exact shape. Because the molecules in a liquid move freely, liquids take the shape of their containers. Molecules in liquid have more space between them than molecules in solids do, but they do not move farther away from each other. A **gas** has no exact shape or volume. Gas particles are also free to move, and they move far away from each other so they can fill their containers. Gas particles usually have large spaces between them.

Solid	Liquid	Gas
Have a definite shape and size. Usually denser than liquids and gases of the same material.	Have a definite size, but do not have a definite shape.	Does not have a definite size or shape, but matches its container.
• Rocks • Ice Cream • Pencils • Apples	• Milk • Water • Juice • Rain	• Steam • Fire • Helium • Fog

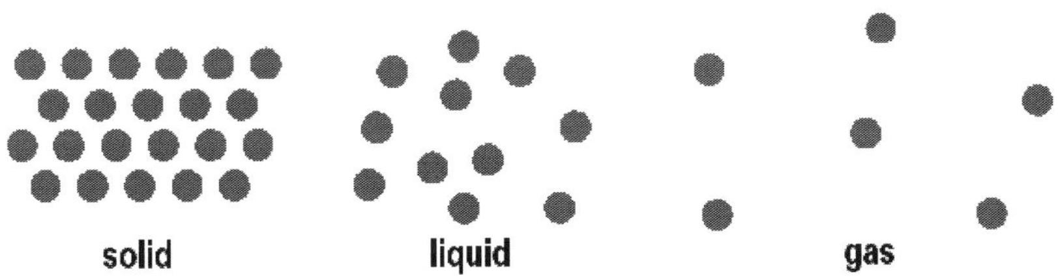

Review Video: States of Matter
Visit mometrix.com/academy and enter code: 742449

Six Different Types of Phase Change

Most materials stay in one state, or **phase**, of matter at normal temperatures. However, when temperatures change, a material may go through a **phase change**. Changing from a solid to a liquid is called **melting**. In contrast, changing from a liquid to a solid is called **freezing**. A liquid changing to a gas is called **evaporation,** or vaporization. A gas changing to a liquid is called **condensation**. In rare cases, solids can change directly to gases, and gases can change directly to solids. A solid changing directly to a gas is known as **sublimation**, and a gas changing directly to a solid is called **deposition**. The temperature of a material must rise for the material to melt, evaporate, or sublimate. The temperature of a material must decrease for the material to freeze, condense, or deposit.

Water is a good reference for freezing, melting, evaporation, and condensation. When liquid water results in conversion to ice. When the ice is heated again, it melts and turns back into liquid water. The melting point for water is 32°F and 0°C. Heating water causes evaporation, resulting in water vapor.The boiling point for water is 212°F and 100°C. When the water vapor cools down enough, it will condense and become liquid water again.

Review Video: Chemical and Physical Properties of Matter
Visit mometrix.com/academy and enter code: 717349

Properties of Matter

Matter is anything that takes up space and has weight. Even air has weight and takes up space. All types of matter have different properties that can be observed or measured. Properties of matter include temperature, mass, density, and magnetism.

Mass and Weight

It is important to remember that mass and weight are not the same. **Mass** is the amount of matter that is present in a material. **Weight** is the **gravitational force** acting upon the mass in a specific gravitational field. A 100-kilogram object has the same mass on Earth as on the Moon because it is the same size and is made of the same materials no matter where it is located. However, the object's weight will differ significantly because the force of gravity is much stronger on Earth than it is on the Moon. An object that weighs 980.7 Newtons (220.5 pounds) on Earth would weigh 163.5 N (36.75 pounds) on the Moon. In space, this object would not have weight, but it would still have a mass of 100 kg.

Temperature

Temperature is a property that tells how hot or cold something is. Temperature also shows the amount of **thermal energy** in an object. For example, when an object is cold and has a low temperature, it has little thermal energy. Sometimes a change in temperature can cause other physical changes. Heat makes many substances expand, while cold makes many objects shrink. Temperature usually transfers from one object to another. For example, when a cup is filled with ice and water, the ice and water exchange thermal energy until they reach the same temperature. This will usually cause the ice to melt.

Volume and Density

Volume is a measurement of the **size** of an object, or how much space an object takes up. Volume affects other properties, such as density. Density is the amount of **mass**, or the amount of matter, in a certain **volume**. More matter within the same volume equals a higher density. **Density** is found using the mass and volume of an object. For example, an inflated balloon contains little mass, but can be large in size. A watermelon is about the same size as an inflated balloon, but it has much more mass. Therefore, the watermelon is denser than the balloon. Density affects whether objects float or sink. For example, think of oil and water. Oil is less dense than water, so if oil and water are placed in the same cup, the oil will **float** to the top. Another way to describe this is to say that the water is actually **sinking** in the oil.

Densities of Solids and Liquids

Density compares the **mass**, or amount of matter, of a material to its **volume**. When an object is changed to a different state of matter, its mass does not change. However, phase changes often result in changes in volume. If a solid sample and a liquid sample have the same volume and are made of the same material, the solid will usually have more matter. This means that the solid sample is denser and will sink in the liquid sample, even though both samples are made of the same material. However, water is different. The molecules in ice are unique, and they move away from each other instead of moving closer together. This means that ice has less matter in it than an amount of liquid water with the same volume Therefore, ice is less dense and floats in water.

Relative Density

Relative density is a special measure of density. **Relative density** is found by comparing the density of one material to the density of another material. The density of water is often used when studing relative density. Density can be used to tell whether one material will float or sink in another material. Many solid objects, such as coins, paperclips, or sunglasses, will sink in water because they have higher densities than water. This means that these objects have higher relative densities than water. Part of the reason these objects have higher relative densities is because they are solid throughout. However, some solid objects are not solid throughout. may be hollow or have one or more openings that allow air inside. These objects, such as ping-pong balls or corks, contain air, which lowers each object's density and causes them to float in water. These objects have lower relative densities when compared to water.

Magnetism

Magnetism is a property that some rocks or metals can possess. **Magnetism** is a force that can push or pull on other magnetic materials. Magnets always have two **poles**, usually called a north pole and a south pole. The **north pole** of a magnet will **attract** the south pole of another magnet and **repel** the north pole. The **south pole** of a magnet will **attract** the north pole of another magnet and **repel** the south pole. The poles on a magnet are usually identified as being north or south because the North and South Poles of the Earth are magnetic. This is the reason that compasses work. The small needle in a compass is attracted to the Earth's North Pole and points in that direction. Many types of technology use magnetism, including electronics, motors, and credit cards.

Magnets

A **magnet** is any object or material, such as iron, steel, or magnetite, that can affect another substance within its field of force that has like characteristics. A **magnetic field** is mapped by invisible curved lines of force that attract magnets or certain metals, such as iron or nickel. A simple bar magnet has a south pole (S) at one end and a north pole (N) at the other end. Each pole will attract the opposite pole of another magnet and repel the same pole. The north pole of a compass needle will point to the south pole of a magnet, while the south pole of a needle will be attracted to the north pole of the magnet.

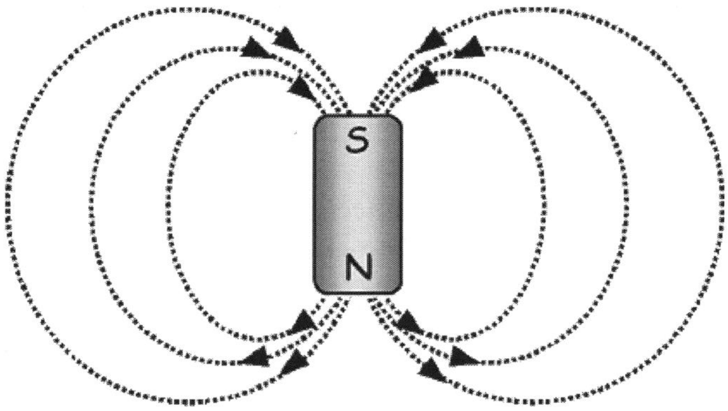

Opposite Poles of Two Magnets Attract Each Other

When two bar magnets are lined end to end with the north and south poles of each near each other, the lines of force run from the north pole of one magnet to the south pole of the other magnet, indicating **attraction**.

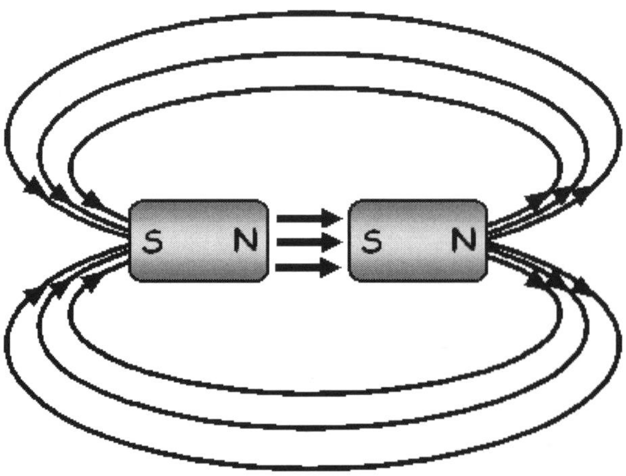

Cutting a Bar Magnet in Two

If a bar magnet is cut into two pieces, two complete magnets will form, each with a north and a south pole. If a bar magnet is cut into three parts, three magnets will form.

Poles of Two Magnets Repel Each Other

When the like poles of two magnets are brought close to each other the lines of force run in opposite directions. This causes the two poles to **push each other away**.

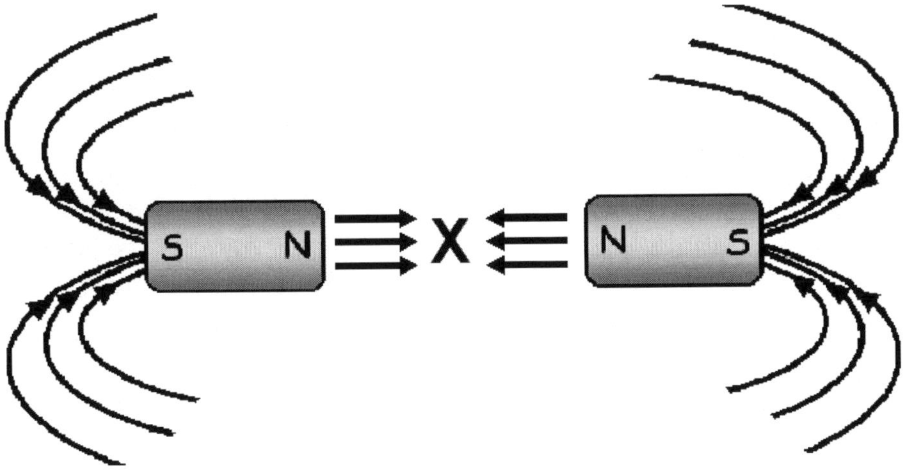

Needle of a Magnetic Compass

The Earth has a magnetic field, similar to a bar magnet. The north pole of a compass needle points to **Earth's North Pole** because the magnetic pole near the North Pole is the south pole of Earth's magnetic field. Likewise, the south pole of the compass needle is attracted to the north pole of Earth's magnetic field near the south geographic pole. Each magnetic pole is attracted to its opposite pole on another magnet or magnetic object.

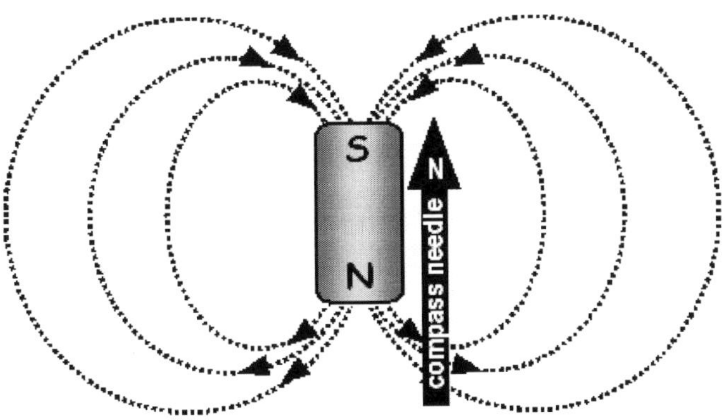

Mixture, Solution, and Colloid

A **mixture** is a combination of two or more substances. A **proportion** is a comparison between the amount of one thing and the amount of another thing. In a mixture, there can be almost any proportion of one substance to another. The proportion of the substances is an important property of the mixture. There are two types of mixtures: homogeneous and heterogeneous. **Homogeneous** means that the properties of the mixture are the same throughout the mixture. **Heterogeneous** means that the properties of the mixture are not the same throughout the mixture.

Solutions and Solubility

A **solution** is a type of homogeneous mixture. Sometimes substances in a mixture can be separated using a tool, such as a filter or centrifuge. Solutions cannot be separated by simply using a tool. Solutions are made by **dissolving** one or more substances into another substance. The substances that are dissolved are called **solutes**. The substance the solutes are dissolved into is called a **solvent**. For example, in a solution of sugar and water, **sugar** is the **solute** and **water** is the **solvent**. Not all materials are able to be dissolved. Materials that cannot be dissolved are **insoluble**, and materials that can be dissolved are **soluble**. Sugar is considered **soluble** in water, but sand does not dissolve in water, so it is considered **insoluble** in water. Most soluble materials are able to be dissolved in water. That is why water is called the **universal solvent.**

A **colloid** is a type of heterogeneous mixture. Colloids are unique because they contain microscopic particles that are suspended in the liquid. This is different from the particles being dissolved. Instead of combining with the liquid, the particles float evenly throughout the liquid. Therefore, these particles can be separated using a centrifuge. A **centrifuge** is a machine that spins small containers at high speeds. This creates a force that moves the particles to the bottom of the container. Milk is an example of a colloid.

Physical Properties of Mixtures and Solutions

Making a mixture or solution does not result in a chemical change to the ingredients. The original ingredients are still present in the mixture or solution. The **physical properties** of each ingredient can usually still be observed in the final mixture. For example, in a solution of salt and water, salt dissolves in the water. However, since the salt is still present, the solution tastes salty. The salt also still has the same mass and volume. The mass and volume of the solution are equal to the mass and volume of the salt plus the mass and volume of the water of the water.

Some properties, such as magnetism, are not taken on by other ingredients in the mixture. For example, if a person mixes sand and iron filings together, then holds a magnet over the mixture, the iron filings would separate from the sand and travel to the magnet, leaving only sand behind.

EXAMPLE

To classify or categorize different materials, scientists conduct tests to measure the materials' physical properties. In which of these scenarios is a scientist investigating the physical properties of a material?

a. A scientist conducts an experiment on the growth of plants in different lighting.
b. A scientist floats two materials in water to determine which is denser.
c. A scientist calculates the effects of friction on the speed of a car.
d. A scientist mixes two chemicals and observes what happens.

Choice A describes an experiment involving plant growth. Plants are living things, and this is an example of observing a process over time rather than an observation of a material's physical properties. Choice A is not the correct answer.

Choice B describes a test used to measure which material is denser. Density is a measurement of how heavy something is given its size. Density is a physical property and is useful for categorizing different materials. Choice B could be the correct answer.

Choice C describes a calculation of factors affecting a car's speed. This is not an example of testing physical properties to classify materials. Choice C is not the answer.

Choice D describes an investigation of what happens when two chemicals are mixed. Physical properties are things like color, magnetism, solubility, and density. Mixing two chemicals doesn't test for any specific physical properties. Choice D is not the correct answer.

Choice B is the only example that describes a test that measures a physical property, which is density in this case. Choice B is the correct answer.

Electricity, Sound, and Light

Electrical Circuits

Electronic devices require electric currents running through them to complete their functions. An **electric current** is the movement of electric charges in one direction. This movement can only follow a closed path. An electric circuit is not complete without a **source of electricity**, **wire** that conducts electricity, and a **device** that uses the electric current to perform a function. The circuit is created by attaching one side of the electricity source, such as a battery, to part of the device using conductive wire. Then, the wire is used to connect another part of the device to the other side of the power source, creating a closed loop. For an electrical circuit to conduct an electric current, the circuit must contain a complete loop made of conductive materials.

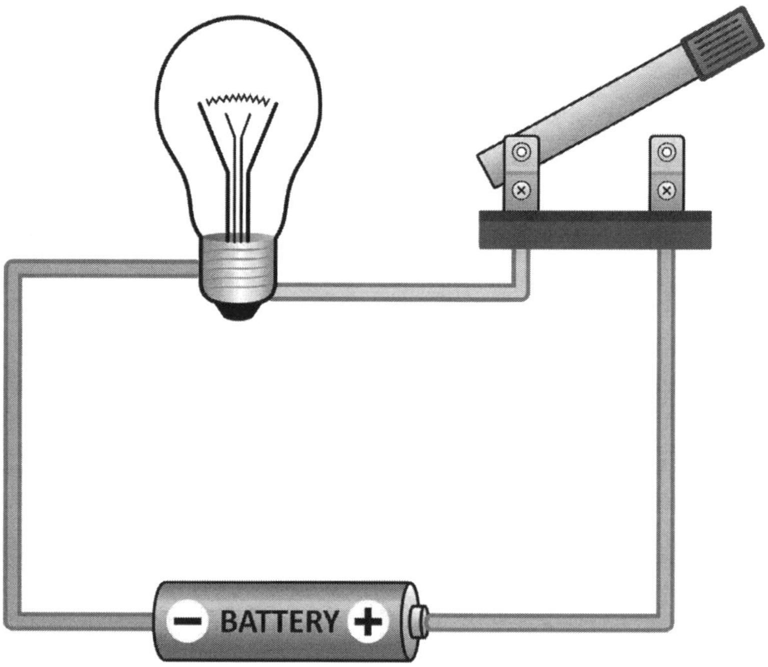

Products of an Electrical Circuit

If the device is a light bulb, a closed circuit will allow the electric current to make a complete loop around the circuit that will enable the light bulb to give off **light**. Similarly, an electric current can produce **sound** when it travels through a buzzer or another device that causes vibration. A **resistor** is an object that can be added to an electric circuit and limits the amount of electricity running through the circuit. As the resistor reduces the electricity, **heat** is produced because the resistor converts the electrical energy into heat. Some appliances use resistors to achieve their main functions. For example, a toaster contains a resistor that it uses to heat food.

Example

What happens when a flashlight is switched on?

a. An open circuit is created, allowing electricity to flow and producing light.
b. A closed circuit is created, allowing electricity to flow and producing light.
c. Stored light energy obtained through the flashlight's solar panels is released.
d. Stored light energy contained within the flashlight's batteries is released.

Choice A describes a circuit, which consists of an electricity source and a path of wire that conducts electricity. When a circuit is open, the wires are not fully connected, which prevents the electricity from flowing through the circuit. If the circuit within a flashlight is open, there is no electricity flowing, and no light is produced. Choice A cannot be the correct answer.

Choice B states that switching on a flashlight creates a closed circuit. A closed circuit is fully connected, which would allow electricity to flow from the flashlight's batteries to the lightbulb, converting electrical energy to light energy. Choice B could be the correct answer.

Choice C has to do with the storage of energy obtained from solar panels, which capture light energy from the Sun. Flashlights use batteries, not solar panels, to power a lightbulb. Choice C is incorrect.

Choice D describes light energy stored within the flashlight's batteries. While batteries do contain energy, it is not in the form of light energy. Instead, the light from a flashlight is created when electricity moves through a circuit, powering a bulb and generating light energy. Choice D cannot be the correct answer.

Choice B is the only answer that accurately describes how flashlights are powered through the completion of a closed circuit, making it the correct answer.

Properties of Light

Light is a type of **energy** that is on the **electromagnetic spectrum**. Light energy travels straight in one direction unless it touches another object that interferes with its path. Therefore, if a flashlight were turned on in space, the light could be seen on the other side of the galaxy, as long as it did not run into an object.

Light and Different Media

When talking about light, a **medium** is a substance through which light travels. There is no air in space, so there is no medium in space. This is why space is sometimes called a **vacuum**. On Earth, air and water are media through which light travels. Even though air is invisible, it slightly interferes with light. Air can cause light to **diffract**, or spread out. A flashlight does not shine in a perfectly straight line due to diffraction in air. Instead, it spreads out and gets weaker with distance from the light source.

Reflection

When light touches an object, it interacts with the object in one of three ways. On shiny objects, such as mirrors, metal, and glass, at least some of the light energy **reflects**, causing it to suddenly change directions. The light hits the object at one angle, then leaves the object at a different angle.

Refraction

When light moves through one medium and then moves through a different medium, the appearance of the light changes because the two media have different **densities**. The density of a medium will affect the properties of a light wave. Those properties control which direction the light travels. Water has a different density than air, so when a light wave enters water from the air, its angle changes. This is called **refraction**. Because human vision is dependent on light, refraction creates an optical illusion when objects are put in water. If you were to insert a pencil into a cup of water, it would appear to bend at the surface of the water. This is because the light waves that allow humans to see the pencil are being refracted.

Absorption

Not all objects reflect light well. When light touches an object that is not shiny, little energy is reflected. The energy is instead **absorbed**. When light is absorbed, the light energy is converted to **thermal energy**, causing the temperature of the object to increase.

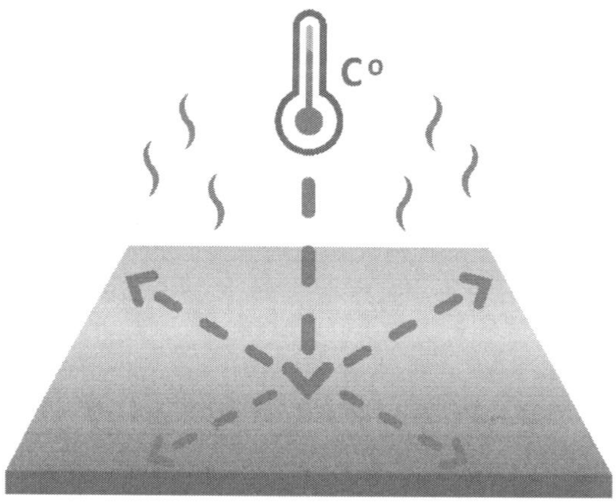

Example

In science class, a student aims a laser pointer at an angled mirror at the back of the classroom. The student observes that the laser beam is reflected at the same angle and notices a red dot on the wall beside her. Which of these statements describes the behavior of light in this scenario?

a. Light particles are carried by the air and disperse throughout a room.
b. The speed of light depends on the density of the surrounding air.
c. When light reaches an object, it moves in a curved line around the object.
d. Light travels in a straight line until it hits an object and changes direction.

Choice A states that light moves in particles that disperse similar to a gas. If this were the case, the light would not reflect off the mirror and would instead disperse in all directions from the laser. Choice A is incorrect.

Choice B states that the speed of light depends on the density of the surrounding air. The speed of light is constant and is not affected by the material it moves through, so this is incorrect.

Choice C proposes that light is able to curve around an object. If this were true, the laser would not reflect off the mirror and would instead curve around it. Choice C is incorrect.

Choice D states that light travels in a straight line, which matches the observation that the light traveled until it hit the mirror and then reflected from the mirror to the other side of the room. It also states that light changes direction when it hits an object like a mirror. Choice D could be the right answer.

Choice D is correct. Light always travels in a straight line until it hits an object where it can be absorbed, reflected, or go through the material.

Conductivity

When a material **conducts** energy, the **energy is transferred** through the material. **Conductivity** is a property that describes how easily a material conducts energy. Conductivity usually involves either heat energy or electrical energy. For example, metals are usually good at conducting both heat and electricity. In contrast, plastics are usually poor conductors of heat and electricity. Materials that conduct energy well are known as **conductors** and materials that do not conduct energy well are known as **insulators**.

Insulators

Insulators do not conduct energy well. They are often used with conductors to **limit** where thermal or electrical energy can flow. For example, frying pans are often made using a conductor and an insulator. Frying pans are made of metal, which conducts heat. However, the handle on a frying pan is usually made of an insulator, such as silicone or plastic, because these materials do not conduct heat well. The heat does not heat up the handle cover, even though the metal inside the cover is very hot. Using an insulator to cover the handle allows the cook to touch the handle even though the metal portion of the pan is hot enough to cook food.

Insulators are also used with conductors in electrical parts. For example, most electrical cords are made of a conductor and an insulator. The wire inside the cord is a conductor that allows electrical energy to flow. The wire is surrounded by an insulator, such as plastic. If a person were to touch a wire that was conducting electricity, the electricity would shock and harm the person. The uncovered wire could also shock a nearby object and cause a fire. The insulator around the wire keeps the energy from flowing outside of the covering. This protects other objects or people from being harmed by electricity in the cord.

Examples of Conductors and Insulators

	Conductors	Insulators
Thermal	Aluminum, copper, silver, diamonds	Plastic, wood, water, air
Electrical	Gold, steel, copper, water,	Rubber, plastic, glass, rock, wood, air, diamond

Sound Waves

The **pitch of a sound** as it reaches one's ear is based on the frequency of the sound waves. A high-pitched sound has a higher frequency than a low-pitched sound. Like all waves, sound waves transmit energy. The rate at which this energy is transmitted is the sonic power. Loudness, or intensity of sound, is the sonic power received per unit area.

When a pair of sound waves with slightly different frequencies interfere with one another causing a periodic variation in sound intensity or a **beat**. The frequency of the variation, called the **beat frequency**, is equal to the difference between frequencies of the two sound waves. The phenomenon is used when tuning two instruments to one another. As the two pitches get closer, the beat frequency will become smaller and smaller until it disappears entirely, indicating that the instruments are in tune.

Review Video: Sound
Visit mometrix.com/academy and enter code: 562378

Living Systems and Ecosystem Interactions

Populations of Animals or Plants

A **population** consists of all of the **organisms** of a **certain kind** in a **defined area**, **region**, or **habitat**. It may be all the red foxes in a given national park, all the loblolly pines in Virginia (very hard to count), all the bullfrogs in a certain pond, or even all of the boxelder bugs on a single box elder tree. In the case of rare or endangered species it may be all of the individuals still living in the wild.

Several factors operate to keep animal and plant populations under control. Predation, grazing, disease, competition for limiting resources such as food or nutrients, water, habitat and living space, hunting and breeding territory, and sunlight for plants all play important roles. Even the size of the population can influence factors such as birth rate and severity of disease outbreaks or force individuals to migrate to other less crowded areas.

Abiotic Factors and Biotic Factors

Abiotic factors are the nonliving physical and chemical factors in the environment. Even though they are not living, the growth and survival of living organisms depend on abiotic factors. These factors can determine the types of plants and animals that will establish themselves and thrive in a particular area. Some abiotic factors, such as oxygen, sunlight, and water, are taken in by organisms. Other abiotic factors are not taken in by organisms, but organisms still need them for shelter and protection from harm. Abiotic factors include:

- Light intensity available for photosynthesis
- Temperature range
- Available moisture
- Types of rocks
- Types of minerals
- Types of atmospheric gases
- Relative acidity (pH) of the system

Biotic factors are the living components of an environment. Biotic factors affect the relationships between organisms and the available resources in an area, which can limit the type and number of resident species. Predator/prey, producer/consumer, and parasite/host relationships can define a community. Biotic factors include:

- Population levels of each species
- Food requirements of each species
- Interactions between species
- Wastes produced

Example

Which of these situations describes an interaction between living and nonliving parts of an ecosystem?

a. A coyote preys upon rabbits
b. A deer eats leaves and berries
c. A mountain lion drinks water from a stream
d. An animal avoids eating a cactus because of its spikes

Choice A is an example of a predator eating its prey. Because the question is asking for an example of an interaction between living and nonliving parts of an ecosystem, one of the factors here needs to be a nonliving part of the animal's physical environment. Coyotes and rabbits are both living things, so this answer can be eliminated.

Choice B is an example of an herbivore eating plants. Plants are living things, so this does not describe a relationship between living and nonliving factors. Choice B is incorrect.

Choice C describes an animal, the mountain lion, drinking water from a stream. Water is a nonliving thing that animals need to survive, so this answer could be correct.

Choice D is an example of a plant protecting itself from an herbivore, in this case with spikes. Both animals and cacti are alive, so this choice can be eliminated.

Choice C describes an example of an animal relying on an aspect of its physical environment, making this an example of an interaction between living and nonliving parts of an ecosystem. Choice C is the correct answer.

Types of Animals

Vertebrate

A *vertebrate* is an animal with an internal *spinal column* (backbone) composed of soft cartilage or hard bone. Vertebrates also have a skull which contains a brain and most have paired *appendages* (legs, flippers or fins). Lampreys, sharks and rays, bony fishes, amphibians (frogs and salamanders), reptiles, birds and mammals are all vertebrates. Image by Ruth Lawson Otago Polytechnic.

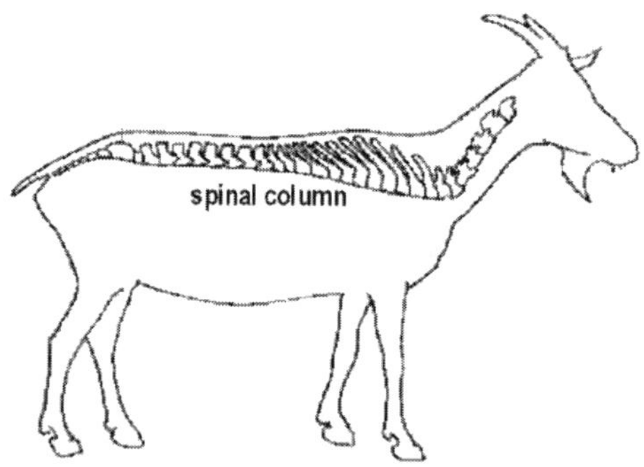

INVERTEBRATE

Invertebrates lack an internal spinal column. Their shape is maintained by a hard, rigid outer shell or covering called an exoskeleton or by the pressure of their internal fluids (like a plastic bag full of water). Insects, spiders, centipedes, snails, clams, octopuses and squids, and earthworms are all invertebrates.

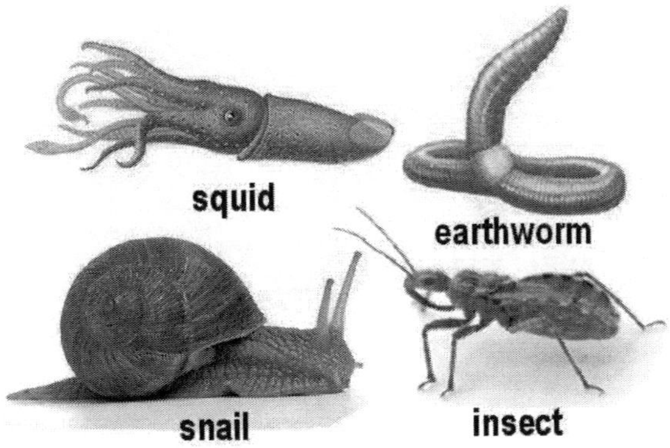

MAMMALS

Mammals are air-breathing, warm-blooded vertebrates with a backbone, skull, internal skeleton and a well-developed brain. Most are covered with hair, and most have specialized teeth. Most give birth to live young, and the mothers produce milk to nourish their young. Birds are also warm-blooded, but they are covered with feathers instead of hair, their young hatch from hard-shelled eggs, and they have no teeth. Bats, platypuses, opossums, kangaroos, elephants, deer, squirrels, apes, monkeys, and whales are all mammals.

REPTILE

Reptiles are air-breathing vertebrates. They cannot generate their own body heat like mammals and birds. Their skin is covered with scales or horny plates. Some lay eggs on land, and others have eggs that hatch inside the mother, and the young are born live. Crocodiles and alligators, turtles, lizards and snakes (like the cobra shown here) are all reptiles. Other reptiles include the dinosaurs, ichthyosaurs, and flying reptiles that died out millions of years ago.

AMPHIBIAN

Amphibians are cold-blooded, air-breathing vertebrates with a smooth and slimy or dry and warty skin. Most lay their eggs in water, but a few lay eggs in wet areas on land. Unlike reptiles and birds, amphibian eggs lack any kind of protective outer shell. Many amphibians undergo metamorphosis from a gill-breathing tadpole stage to an air-breathing adult. Frogs, toads, and salamanders are all amphibians.

FROGS

Frogs are *amphibians*. That means they must lay their eggs in water. The egg cell divides to form an *embryo* that grows until it begins to look like a *tadpole* (Fig. 1). Eventually the tadpole swims freely in the water and breathes with gills. As the tadpole develops it grows one hind leg (Fig. 2), then another (Fig. 3), and finally 2 front legs (Fig. 4). The tail shrinks (Fig. 5) until it disappears and the tadpole has *metamorphosed* into a tiny *frog*.

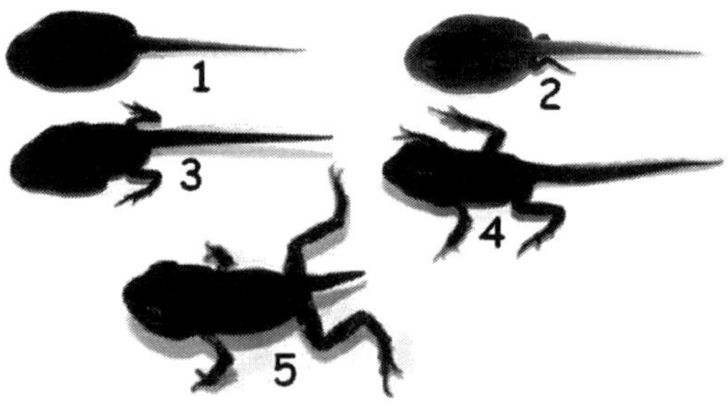

To a scientist there is very little difference between a frog and a toad. All toads are really just frogs. But for most people, a frog usually refers to a tailless amphibian that lives entirely or mostly in or around water and has smooth slimy skin. A toad, on the other hand, has dry, warty skin and lives on land away from water. But both lay their eggs in water, and their tadpoles develop in water.

Fish

Except for lungfish, fishes are gill-breathing animals with a backbone, skull and internal skeleton that live in the water (although some can come out on land for short periods). They have no legs, but most have paired fins for swimming. Most are cold-blooded, although a few fast swimmers can raise their body temperature slightly. They lay their eggs in the water. Sharks, bass, tuna, salmon, goldfish, and catfish are all different types of fishes.

Shark, Dolphin, and Porpoise

Sharks are a type of fish and breathe by means of gills. Although they look somewhat the same, dolphins and porpoises are mammals that live in the ocean. They breathe by means of lungs, and they must come to the surface to take in air through blowholes on top of their heads. Sharks swim by moving their vertical tails from side to side. Dolphins move their horizontal flukes up and down.

Insects

Insects are *arthropods* (joint-legged animals) whose bodies are divided into three parts: a head, thorax, and abdomen. Insects all have a hard outer covering called an *exoskeleton*, three pairs of legs (6 in all), compound eyes, and a single pair of antennae. They have to molt and shed their hard outer skin in order to grow. More than half of all living animals are insects. Spiders, scorpions, centipedes and millipedes are not insects but belong to different groups of arthropods.

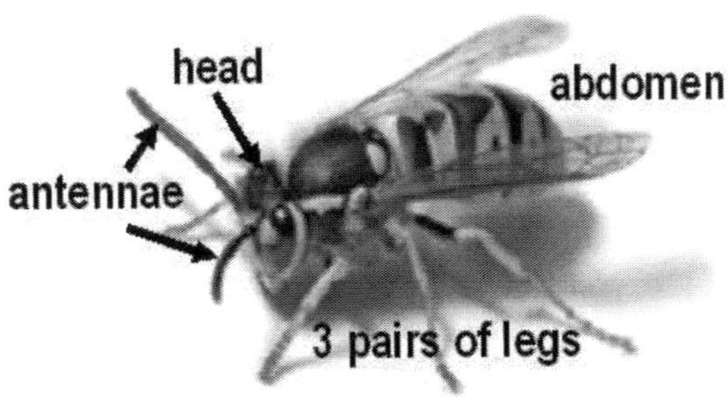

Fireflies

Fireflies or "lightning bugs" are actually flying beetles that come out at night. They have special light producing organs in their bellies that create a cold form of biological light. They flash their light on and off in a pattern unique to each species in order to attract a mate. Females of some predatory species of firefly imitate the flashing pattern of the females of a different species to attract the male. When the male of the other species approaches, she kills and eats him.

Spiders and Scorpions

Spiders and scorpions are not insects, but arachnids. Spiders and scorpions have four pairs of walking legs, simple (not compound) eyes, and no antennae. The body is divided into two parts: a cephalothorax (the fused head and thorax) and an abdomen.

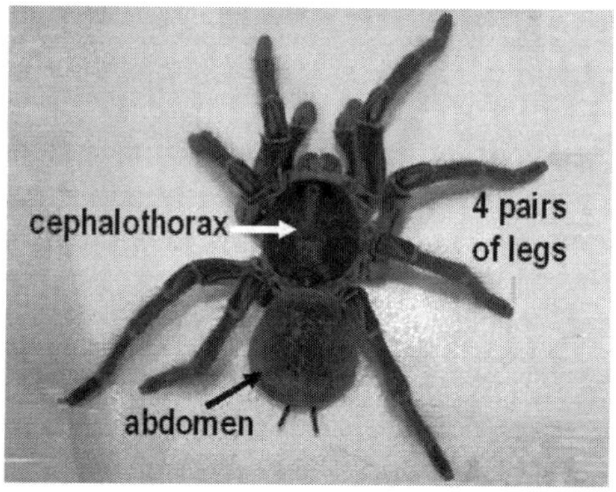

Spider Webs

Most spiders produce silk from glands at the tip of their abdomen. These glands can produce a trailing safety line and sticky silk making a web to trap their prey. An orb web like the one shown here lets the spider sit and wait for prey to be caught in the sticky web. This web helps the spider save energy since it does not have to hunt and chase down its prey like some spiders do. Weight for weight, spider silk is stronger than steel.

Millipede

Millipedes and centipedes are not insects. Millipedes and insects are both arthropods, but that is where the similarities end. Millipedes are called *myriapods*. Millipedes have 1 pair of antennae, but only simple eyes called *ocelli*. They have a head, a great many body segments, and *two pairs of legs* on most segments. Most eat dead plant matter. Insects have a pair of antennae, but they have compound eyes, a head, thorax, and abdomen, and just three pairs of legs.

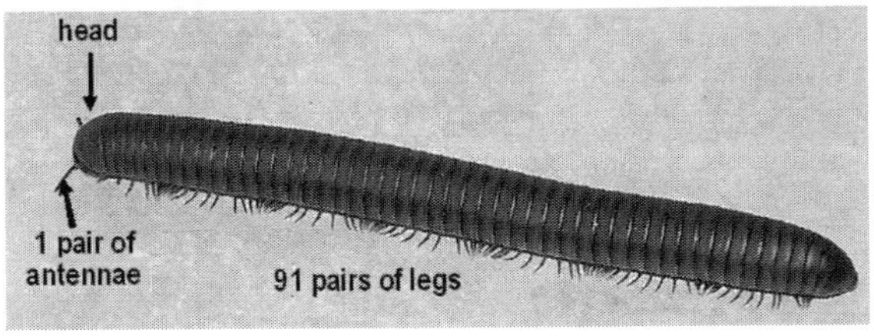

Centipedes are called *chilopods*. They have a pair of antennae, many body segments, and *a single pair of legs* on each segment. Centipedes are carnivorous and have a pair of fangs formed from their first legs that they use to inject venom into their prey. Insects have a pair of antennae, but they have compound eyes, a head, thorax, and abdomen, and just three pairs of legs.

Herbivores, Carnivores, and Omnivores

Herbivores are animals that only eat plants. Examples are rabbits, deer, antelope, rhinoceroses and elephants. *Carnivores* are animals that only eat other animals, like lions, wolves, foxes, most bats, hawks and owls, insect-eating birds, snakes, and sharks. *Omnivores* are animals that eat both plants and other animals. Examples include bears, pigs, skunks, raccoons, birds like crows and ravens, and human beings.

Vultures

Vultures are mostly *scavengers* that eat *carrion*— the dead and decaying flesh of animals. When an animal dies it immediately begins to decay or rot. Bacteria break down the dead tissues and release a foul smell that attracts scavengers like vultures, hyenas, crabs and lobsters, and certain flies that feed on decaying flesh. Many carnivores like coyotes, hyenas, opossums, hawks and eagles will also eat carrion in addition to catching live prey.

Predator and Parasite

Predators are animals that attack and directly kill other animals for food. Lions, foxes, bats, snakes, hawks, and owls are examples of predators. *Parasites* are organisms that live on or inside another organism called a *host*. The parasite drains energy from the host and may release substances that harm it. Many parasites cause the host t to get sick and eventually may even kill it. However, with parasites the host is not killed outright as with predation. Bacteria, viruses, certain kinds of worms, mosquitoes, and ticks are examples of parasites.

Producers, Consumers, and Decomposers

Producers are organisms that can make their own food. Most producers are plants. Plants use sunlight, carbon dioxide from the air, and water to make sugars that provide **energy** through a process called **photosythesis**. Plants only need sunlight, water, and certain minerals and other nutrients to live, grow, and reproduce. **Consumers** are organisms that eat other organisms. **Decomposers** are organisms that feed on decaying plant and animal matter. Since decomposers cannot make their own food, they are classified as consumers. Fungi, such as mushrooms, are decomposers. They break down the tissues and wood of living or dead plants or the bodies of dead animals.

Energy Pyramid

Energy flows through an ecosystem as producers and consumers are consumed by other organisms. The flow of energy can be tracked through an energy pyramid. An **energy pyramid** shows how energy is transferred from one trophic level to another. **Trophic levels** describe whether an organism is a producer or a consumer and how much energy is typically passed to them when they eat. Producers always form the **base** of an energy pyramid, and the energy they pass along comes from the **Sun**. Consumers form successive levels above the producers. Producers only store about 1% of the solar energy they receive. Then, each successive level only uses about 10% of the energy of the previous level. Therefore, **primary consumers** use about 10% of the energy used by primary producers, such as grasses and trees. Next, **secondary consumers** use 10% of the energy used by primary consumers, or 1% overall. This continues up for as many trophic levels as exist in a particular ecosystem.

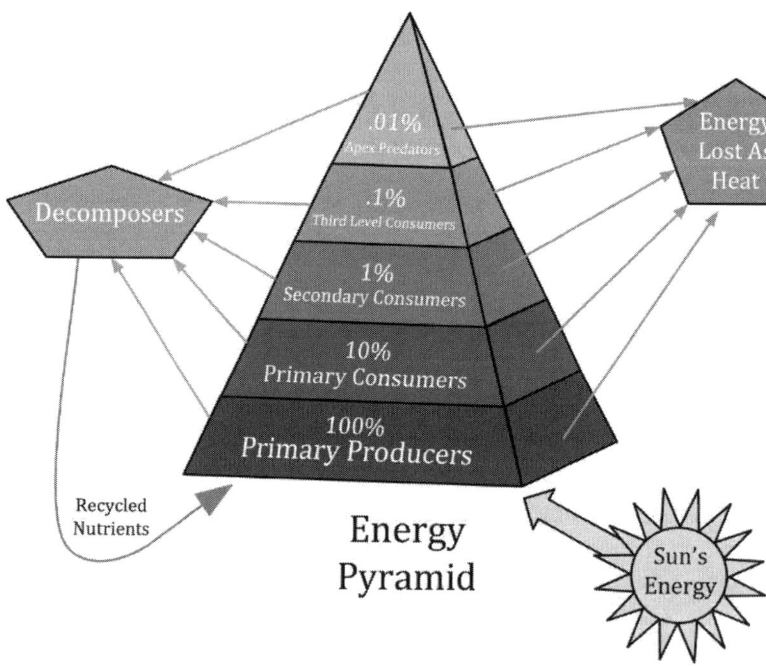

Food Web

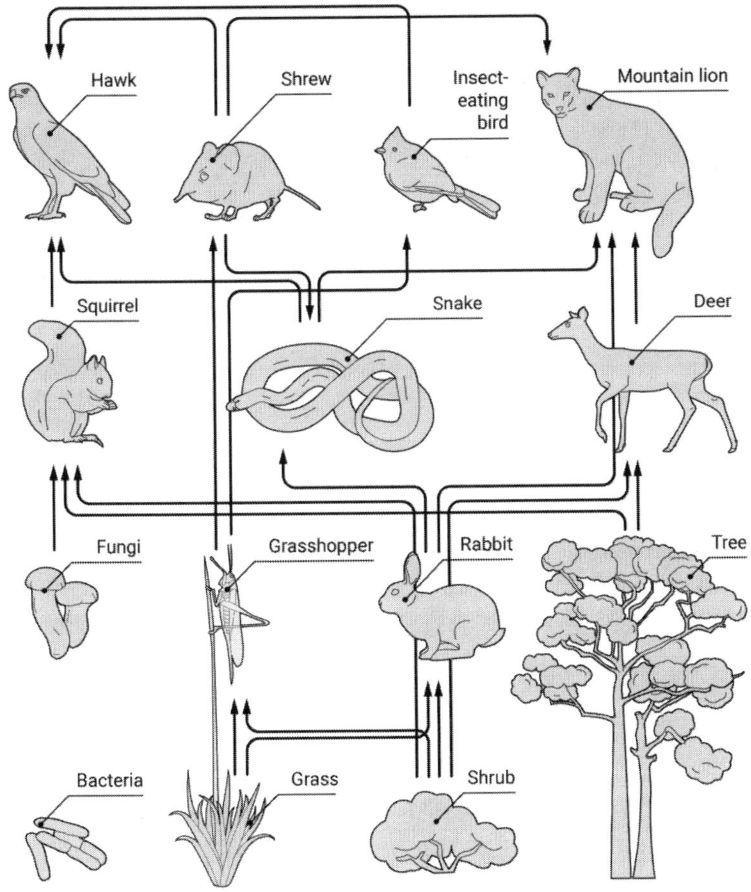

Energy flow through an ecosystem can also be illustrated by a **food web**. Energy moves through the food web in the direction of the arrows. In the food web above, **producers** such as grass, trees, and shrubs use energy from the Sun to produce food through photosynthesis. **Herbivores** or **primary consumers** such as squirrels, grasshoppers, and rabbits obtain energy by eating producers. **Secondary consumers**, which are carnivores such as snakes and shrews, obtain energy by eating the primary consumers. **Tertiary consumers**, which are carnivores such as hawks and mountain lions, obtain energy by eating the secondary consumers.

Review Video: Food Webs
Visit mometrix.com/academy and enter code: 853254

Interdependence of the Food Web

Because each level of consumer is dependent on the previous level for food, the population of each level affects the other animal groups. For instance, if an ecosystem is made up of only grass, deer, and wolves, grass is the producer, the deer eat the grass, and the wolves eat the deer. If deer are overhunted one year, the grass is given more room to grow because less of it is consumed. However, the wolves will not have enough food, so the wolf population will reduce in size. Eventually, because of the abundance of grass and a reduced population of wolves, the deer may then have a surge in population the next year. This example is not a realistic example, as there are usually many more producers, consumers, and predators within an ecosystem.

Example

Which of the following accurately depicts the first three steps in the movement of energy through an ecosystem?

a. Sun → producers → consumers
b. Producers → Sun → consumers
c. Consumers → producers → decomposers
d. Decomposers → consumers → producers

Choice A depicts the movement of energy from the Sun to producers and then to consumers. Producers get energy from the Sun through photosynthesis, while consumers get energy by eating plants or other animals. The path depicted in Choice A matches this pattern, so Choice A could be the correct answer.

Choice B depicts energy moving from producers to the Sun and then to consumers. Producers such as plants gain energy from the Sun, so the Sun should come first in this pathway. Choice B is incorrect.

Choice C describes the movement of energy from consumers to producers and then to decomposers. Producers get energy from the Sun while consumers eat other living organisms, so producers should always come before consumers in an energy pathway. Choice C is not the answer.

Choice D depicts the movement of energy from decomposers to consumers and then to producers. Decomposers get energy by breaking down the remains of plants and animals that have died, recycling nutrients into the ecosystem. Decomposers cannot be the first step in energy transfer, so Choice D is incorrect.

Choice A is the correct answer because it accurately shows the movement of energy from the Sun to producers through photosynthesis, followed by consumers that eat other living organisms.

Relationships Between Organisms

Organisms can interact with one another in many ways. For example, **herbivores** eat plants, and **predators**, or **carnivores**, eat other animals. Organisms can also **compete** for biotic or abiotic resources in an ecosystem. In addition, organisms can rely on each other through **mutualism**, which is a relationship between at least two organisms in which both organisms benefit from each other. One organism can also rely on another through **commensalism**, which is a relationship between at least two organisms in which one organism benefits while the other is not harmed or benefitted.

Competition between Different Organisms

Animals and plants have to compete with other species for food or nutrients, water, places to live, nesting or breeding sites, sunlight in the case of plants, and other factors in the environment that may be scarce or limited. In addition, animals and plants of the same species have to compete with each other for these resources and for the opportunity to breed and reproduce. Competition may seem harmful to some plants and animals, but it actually helps some species. Competition helps control population sizes and prevents **overpopulation**. Overpopulation of a species can be dangerous since there may not be enough resources to support all of the animals of that species or animals of other species.

Changes to Ecosystems

Most ecosystems rely on **consistent patterns** and a **balance** between organisms and resources. When something changes this balance, the ecosystem may suffer. For example, consider a grassland ecosystem. Deer are grazers, animals that eat grass, and they have natural predators, such as wolves. If something happens and the predators of the deer disappear, the ecosystem may suffer. The deer would likely become overpopulated because fewer deer would be hunted by the predators. This would become a problem because as the population grows, there may not be enough grass in the area to feed all of the deer and all of the other animals that eat grass. As a result, grass may become scarce and other animals, such as insects, would lose their habitats since there would be little grass. This situation shows that even one change can hurt many members of an ecosystem.

Humans can also harm ecosystems. For example, when a new road or building is created, it takes up land that had grass, trees, bushes, or other plants growing on it. These plants provide shelter, food, and fresh oxygen for many animals. When these resources are removed, the animals must find them elsewhere. This not only harms the original ecosystem, but it can also harm the ecosystems the animals move to. The animals can cause too much competition in the new ecosystems if there are not enough resources to provide for the new animals and the animals that were already there. If this occurs, organisms from both the old and new populations could suffer.

Traits and Environments

Each organism has **traits** that affect that organism's ability to **survive**. Some of these traits are common among different species that live in similar environments. For example, many desert animals have traits that help them either survive on a small amount of water or extract water from their surroundings. A trait may also help an organism obtain food or shelter in its environment. For example, flamingos have beaks that are shaped in a way that allows flamingos to find food underwater. Furthermore, flamingos have webbed feet so they can easily live in shallow water. Woodpeckers have beaks that are shaped in a way that allows them to find food under tree bark. In addition, they have talons so they can hold onto tree limbs and build nests. These traits help both species of bird survive in their environments.

Traits for Warding Off Predators

Many organisms have traits that protect them from predators. For example, an animal's hair color or skin color may help it blend in with its surroundings and hide. Some animals have skin that can hurt or taste bad to predators, which can lead predators to avoid trying to eat that animal. Many of these traits are the result of **adaptations**, or traits that species developed over time to help them live in their environment.

Example

Which traits would make a species more successful in an aquatic environment?

a. A beak
b. Webbed feet
c. A tail
d. Opposable thumbs

Choice A suggests a beak, which could be useful in obtaining certain types of food. A trait that would make a species more successful in an aquatic environment should be specifically related to living in water. A beak would probably be more useful on land, so Choice A is incorrect.

Choice B suggests webbed feet. Webbed feet could be useful in an aquatic environment because the extra skin between the animal's toes could be helpful for swimming. Additionally, webbed feet are unlikely to be helpful in non-aquatic environments. Choice B could be the correct answer.

Choice C suggests a tail, which is common in animals that live in many habitats, not just in the water. Choice C is not the correct answer.

Choice D suggests opposable thumbs such as those found in humans and other primates. Opposable thumbs allow animals to use tools and do things with their hands, which would be more helpful on land than in the water. Choice D is incorrect.

Choice B, webbed feet, is the only trait that is specifically related to an animal's success in an aquatic environment, making it the correct answer.

Biome

A **biome** is a region that includes the plants and animals best suited to the soil type and climate in an area. This concept is related to the relationships between organisms and their environment. Since plants are more abundant than animals, biomes are often described based on the type of vegetation that exists in the older ecosystems within the biome. Though the word *biome* only refers to the living things in an area, each type of biome is associated with a certain climate. When looking at where the different types of biomes are located around the world, a pattern can be observed. A biome type may appear on different continents, but across those continents, the areas where that biome appears will be located at similar **latitudes**. This is because each type of biome is closely related to certain climate and soil types, and similar climates and soil types tend to appear at similar latitudes.

Savanna Biome

The **savanna biome** is often found near the edge of an **equatorial rainforest**, which is a type of rainforest that exists near the equator. The savanna biome is commonly found in regions that experience a wet-dry tropical climate. During the dry season, the lack of moisture in the soil causes trees in the savanna biome to grow with much space between them. This allows a dense lower layer of grasses and other plants to grow. Though fire is fairly common in savannas during the dry season, the vegetation in these regions tends to be quite fire resistant. In fact, many geographers believe that such fires prevent rainforest vegetation from overrunning the savanna biome.

Grassland Biome

The **grassland biome** is commonly subdivided into two types: tall-grass prairie and steppe. The **tall-grass prairie** type of grassland does not have trees and has much tall grass and broad-leaved herbs called forbs. This kind of biome usually forms in regions with distinct summer and winter seasons, in subtropical or midlatitude climates. The **steppe, or short-grass prairie,** type of grassland is known for having scattered clumps of short grasses and the occasional shrub or small tree. Many grass species and forbs populate this kind of region. Steppes are usually found in midlatitude areas with dry continental climates. In steppes, vegetation grows in wet spring months and becomes dormant during dry summer months.

Desert Biome

The **desert biome** may be subdivided into two types: semidesert and dry desert. The **semidesert** type of desert is found at a wide range of latitudes. It is characterized by thinly-spaced shrubs that have adapted to survive in arid climates. **Steppe regions** may be converted to semidesert through events such as a high number of cattle moving through the area, which tread upon and consume regional vegetation. The thorn-tree semidesert is found in the tropical zone, and has a long, dry hot season and a short, severe rainy season. Thorny vegetation (such as cacti), called thorn-bush or thorn-woods, populate the thorn-tree semidesert.

The **dry desert subdivision** is even barer than the semidesert. Only tough plants that have adapted to live in the climate, such as cacti and hard grasses, can survive under desert conditions. Many dry desert areas display no plant life at all. This is due to the sand that covers the ground, which does not allow for conditions that help plants grow.

Tundra Biome

The **tundra biome** may be subdivided into two types: arctic or alpine. The **arctic tundra** exists at very high latitudes near the poles. During the brief summer season in these areas, low vegetation such as herbs, mosses, and grasses are able to grow because above-freezing air temperatures allow the surface layer to melt. The type and number of plant species depends on the moisture levels in the (usually frozen) ground during the warm season. In the cold season, frost action snaps roots in the ground. Freezing wind and snow kill plants above the ground. **Alpine tundra** is located at high elevations, above the tree line and below bare mountain tops. Physically, the alpine tundra is quite similar to the arctic tundra.

Forest Biomes

The **forest biome** has an abundance of soil and moisture and has warm temperatures for a portion of the year. Trees are the dominant vegetation type in this biome. There are many different types of forest biomes:

- **Low-latitude rainforests** exist in equatorial and tropical climate zones. They contain many different species of tall, densely-spaced trees, normally with broad leaves. The canopy created by this vegetation casts shade on the ground below, which limits the growth of plants on the bottom layer.
- **Monsoon forests** may also be found in tropical latitude zones. However, this type of forest exists in regions that experience a wet season and a dry season. During these seasons, many tree species lose their leaves. This allows the vegetation on the lower layer of the forest to grow.
- **Subtropical evergreen forests** have fewer, shorter tree species than low-latitude rainforests. They also include a lush lower layer of vegetation.
- **Midlatitude deciduous forests** are common in regions of the northern hemisphere that have distinct seasons. During the warm summer season, the tall trees in this type of forest form a nearly closed canopy. In the winter season, these trees lose their leaves. The lower level vegetation also varies with the season, as they grow in the spring and die out when the dense upper canopy forms in the summertime.
- **Needleleaf forests** are made up largely of conifers. Many trees in this type of forest are evergreen, meaning that they shed their leaves once every several years. The spacing of the trees in this forest class may be dense, preventing the growth of vegetation on the ground.

Organisms and Environmental Changes

Sudden environmental changes can include **natural disasters**, **diseases**, and **habitat destruction**. Examples of more **gradual** changes include **climate change** or **gradual habitat loss** due to population growth, agriculture, or the harvesting of natural resources. Both sudden and gradual environmental changes have the potential to significantly disrupt the balance of an ecosystem. When organisms die due to natural disasters, disease, or environmental changes, the number of individuals in a population changes. In more sudden environmental changes, these shifts in population numbers are more dramatic. Additionally, these changes in population numbers can have a **domino effect** on the ecosystem. For example, when a predator population decreases, the prey populations increase because of the reduced risk of being hunted. However, another predator population could increase because there is now reduced **competition** for resources. Environmental changes can also affect populations by influencing the presence of certain physical traits. Because individuals with certain traits are more likely to survive specific environmental changes, those traits will become more common in a population through the process of **natural selection.**

Example

A remote island consisting of rainforest habitats has experienced consistently rising temperatures over the course of a decade. Scientists have made several observations of changes to local populations. Which of the following observations is most likely a beneficial change to a trait that will affect subsequent generations?

a. The population of frogs has declined.
b. Old-growth trees are producing fewer leaves.
c. Birds spend less time in the canopy.
d. A small mammal species produces a thinner coat of fur.

Choice A describes a decline in the population of frogs. Rising temperatures are likely to cause certain populations to decline, particularly amphibians. However, the question asks for an example of a beneficial change to a trait that will be inherited by subsequent populations. In this case, population decline is not a change to a trait, and population decline is rarely beneficial. Choice A is not the answer.

Choice B describes trees producing fewer leaves. This answer specifies that the trees are old-growth trees, meaning they have been around for a long time. In this case, producing fewer leaves is likely to be an individual response to limited resources and environmental changes. This is unlikely to be a beneficial change because fewer leaves mean the trees cannot photosynthesize as much. In addition, this describes an immediate survival response rather than an inherited change to a trait. Choice B is probably not the answer.

Choice C describes a change in birds' behavior, who begin to spend less time in the canopy, likely to avoid the sun. Because this is a change in behavior, it is unclear whether it would be passed on to subsequent generations. Choice C is probably not the answer.

Choice D describes a change to a physical trait, specifically fur thickness in small mammal species. With rising temperatures, having thinner fur is likely to be beneficial in keeping the animal from overheating. Additionally, a physical trait is more likely to be passed on to subsequent generations than a behavioral change. Choice D is the correct answer.

Choice D is the correct answer because it describes a change to a trait that is most likely to be beneficial and to be inherited by subsequent generations.

Human Impact on Marine Environments

Humans can impact ecosystems directly and indirectly. For example, humans can **directly** impact an ecosystem through deforestation and overfishing, while humans can **indirectly** impact ecosystems through the burning of fossil fuels, which can contribute to climate change and pollution. Marine ecosystems are some of the most vulnerable to these impacts. For example, the ocean absorbs some of the carbon dioxide (CO_2) that goes into the atmosphere, and so as more CO_2 goes into the atmosphere due to human activity, more CO_2 goes into the ocean. This absorption has caused chemical reactions that have led the oceans to become more acidic. This **ocean acidification** poses a threat to shell-building organisms and coral. Because coral reefs provide food and habitats for a significant portion of marine life, ocean acidification poses intense danger for marine ecosystems. Human agricultural activity also contributes to ocean pollution, as **runoff** from fertilizers and other chemicals leaks into the ocean.

Overfishing and Artificial Reefs

Reliance on oceans for food has led to **overfishing**, which is when fish are caught and removed from a body of water at a higher rate than the fish can reproduce and replenish. This threatens fish populations as well as entire marine ecosystems. To increase fish populations in a certain area, humans have used **artificial reefs**, which are created by submerging materials into the water to replicate a coral reef habitat. Although artificial reefs can quickly increase the populations of fish and other marine life, they can have negative impacts, such as contributing to ocean pollution. Ironically, they can increase the fish population so much that many people will fish there and repeat the process of overfishing. Because of this, overfishing is a cyclical process that must constantly be kept in check.

Effects of Temperature, Light, Moisture, and Air on Plant Growth

Environmental factors including temperature, light, moisture, and air can all impact plant growth. **Temperature** is the most important environmental factor determining plant growth. The temperature needs vary for different plants. Tender plants will not grow in cold conditions. Plants use photosynthesis to create food. For photosynthesis to take place, plants need **light**. Without light, plants cannot produce food and will die. Water is essential for all life, including plants. **Water** is the main component of plant cells and helps keep the plant turgid. A plant without water will wilt and die. Water is required in photosynthesis and is used to transport nutrients in the plant. **Air** is required for both the production of food (photosynthesis) and the breakdown of food into energy (respiration).

Reproduction, Development, and Growth in Animals

Gamete Formation

Gametogenesis is the formation of gametes, or reproductive cells. Gametes are produced by meiosis. **Meiosis** is a special type of cell division that consists of two consecutive mitotic divisions referred to as meiosis I and meiosis II. **Meiosis I** is a reduction division in which a diploid cell is reduced to two haploid daughter cells that contain only one of each pair of homologous chromosomes. During **meiosis II**, those haploid cells are further divided to form four haploid cells. **Spermatogenesis** in males produces four viable sperm cells from each complete cycle of meiosis. **Oogenesis** produces four daughter cells, but only one is a viable egg and the other three are polar bodies.

Fertilization

Fertilization is the union of a sperm cell and an egg cell to produce a zygote. Many sperm may bind to an egg, but only one joins with the egg and injects its nuclei into the egg. Fertilization can be external or internal. **External fertilization** takes place outside of the female's body. For example, many fish, amphibians, crustaceans, mollusks, and corals reproduce externally by **spawning** or releasing gametes into the water simultaneously or right after each other. Reptiles and birds reproduce by **internal fertilization**. All mammals except monotremes (e.g. platypus) reproduce by internal fertilization.

Embryonic Development

Embryonic development in animals is typically divided into four stages: cleavage, patterning, differentiation, and growth. **Cleavage** occurs immediately after fertilization when the large single-celled zygote immediately begins to divide into smaller and smaller cells without an increase in mass. A hollow ball of cells forms a blastula. Next, during patterning, gastrulation occurs. During gastrulation, the cells are organized into three primary germ layers: ectoderm, mesoderm, and endoderm. Then, the cells in these layers differentiate into special tissues and organs. For example, the nervous system develops from the ectoderm. The muscular system develops from the mesoderm. Much of the digestive system develops from the endoderm. The final stage of embryonic development is growth and further tissue specialization. The embryo continues to grow until ready for hatching or birth.

Postnatal Growth

Postnatal growth occurs from hatching or birth until death. The length of the postnatal growth depends on the species. Elephants can live 70 years, but mice only about 4 years. Right after animals are hatched or born, they go through a period of rapid growth and development. In vertebrates, bones lengthen, muscles grow in bulk, and fat is deposited. At maturity, bones stop growing in length, but bones can grow in width and repair themselves throughout the animal's lifetime, and muscle deposition slows down. Fat cells continue to increase and decrease in size throughout the animal's life. Growth is controlled by genetics but is also influenced by nutrition and disease. Most animals are sexually mature in less than two years and can produce offspring.

Plant Life Cycles and Reproductive Strategies

Pollination Strategies

Pollination is the transfer of pollen from the anther of the stamen to the stigma of the pistil on the same plant or on a different plant. Pollinators can be either **abiotic** (not derived from a living organism) or **biotic** (derived from a living organism). Abiotic pollinators include wind and water. Approximately 20% of pollination occurs by abiotic pollinators. For example, grasses are typically pollinated by wind, and aquatic plants are typically pollinated by water. Biotic pollinators include insects, birds, mammals, and occasionally reptiles. Most biotic pollinators are insects. Many plants have colored petals and strong scents, which attract insects. Pollen rubs off on the insects and is transferred as they move from plant to plant.

Seed Dispersal Methods

Methods of **seed dispersal** can be abiotic or biotic. Methods of seed dispersal include gravity, wind, water, and animals. Some plants produce seeds in fruits that get eaten by animals and then are distributed to new locations in the animals' waste. Some seeds (e.g. dandelions) have structures to aid in dispersal by wind. Some seeds have barbs that get caught in animal hair or bird feathers and are then carried to new locations by the animals. Interestingly, some animals bury seeds for food storage but do not return for the seeds. The seeds of aquatic plants can be dispersed by water, while the seeds of plants near rivers, streams, lakes, and beaches (e.g. coconuts) are also often dispersed by water. Some plants, in a method called **mechanical dispersal**, can propel or shoot their seeds away from them even up to several feet. For example, touch-me-nots and violets utilize mechanical dispersal.

Alternation of Generations

Alternation of generations, also referred to as **metagenesis**, contains both a sexual phase and an asexual phase in the life cycle of the plant. Mosses and ferns reproduce by alternation of generations: the sexual phase is called the **gametophyte**, and the asexual phase is called the **sporophyte**. During the sexual phase, a sperm fertilizes an egg to form a zygote. By mitosis, the zygote develops into the sporophyte. The sporangia in the sori of the sporophyte produce the spores through meiosis. The spores germinate and by mitosis produce the gametophyte.

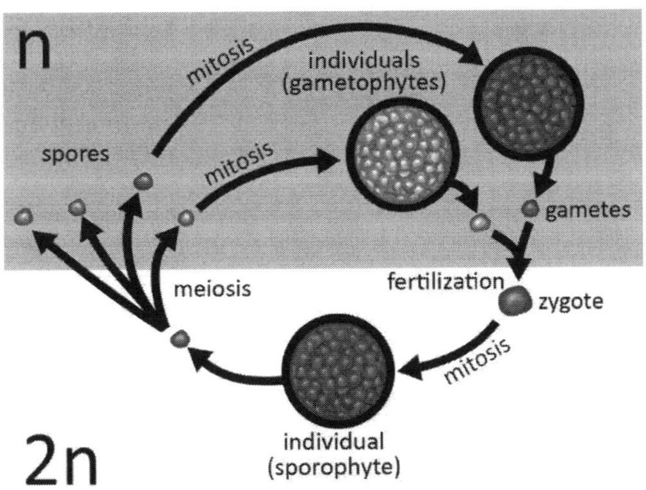

Major Plant Processes

There are three key processes that plants need to survive: photosynthesis, respiration, and transpiration. **Photosynthesis** is the process that plants use to create food. Photosynthesis requires water, carbon dioxide, and sunlight. Photosynthesis creates glucose molecules that can be broken down and used as energy for the plant. The cellular **respiration** process breaks glucose molecules into usable energy for the cell. Respiration requires oxygen to take place, so the plants take in oxygen. **Transpiration** is the process of plants losing water through their leaves. Plants can control the amount of water lost through their leaves. Water is lost through stomata on the underside of the leaves. Stomata can be opened and closed by guard cells.

Changes Throughout the Life Cycle

Many types of plants and animals go through a process of changes throughout their lifespans. This process involves a complete change in how the plant or animal looks and acts during that life stage. In insects and in amphibians, this is called **metamorphosis**.

Plants

A typical plant has three major sections; roots, stems, and leaves. The *roots* lie underground where they absorb water and minerals needed by the plant. These water and minerals are carried up to the leaves through special tubes in the *stem* or *trunk*. The *leaves* use the water from the roots and absorb carbon dioxide from the air. The green color in *leaves* is due to *chlorophyll*. Chlorophyll absorbs sunlight to produce a special chemical that releases energy to enable the plant to produce sugar to nourish the rest of the plant. This food-making process is called *photosynthesis*.

FLOWERS

Flowers are the structures which many plants reproduce. A typical flower has both male and female parts. The male *stamens* consist of a stalk called a *filament* with a swollen *anther* at the tip which produces the *pollen*. The female parts consist of an *ovary* which produces the seeds and a tube called a *pistil* through which the pollen enters to fertilize the seeds. Showy colorful *petals* surround the male and female parts. Below the petals are smaller *sepals* that are usually green.

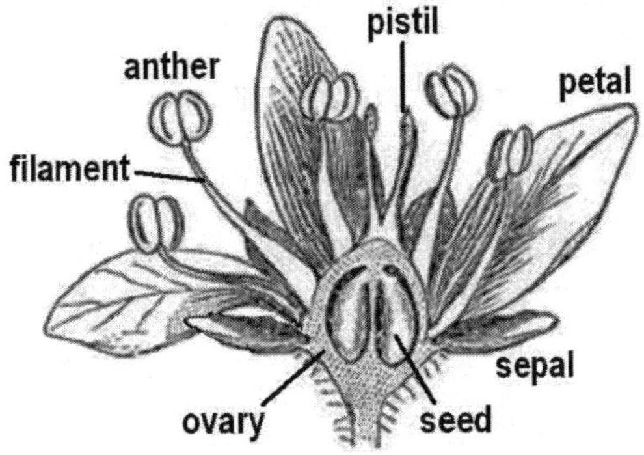

FOOD MANUFACTURING OF GREEN PLANTS

Green plants use energy from sunlight and carbon dioxide from the air to make sugars and energy-carrying molecules by *photosynthesis*. This word means "to put together with light." In this process plants give off oxygen as a waste product. This oxygen is then breathed in and used by animals.

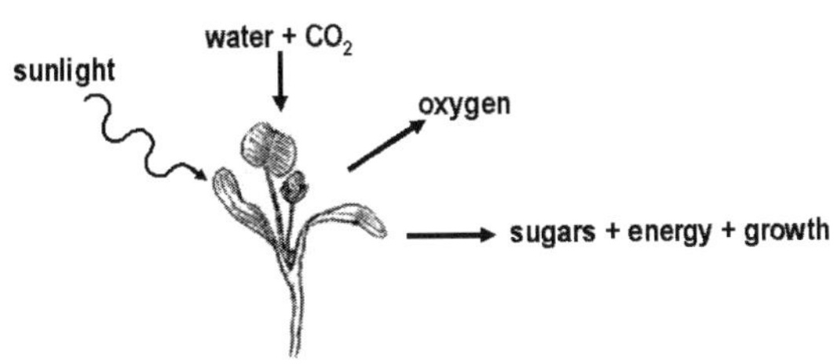

POLLINATION

Pollen can be carried to the female parts of other plants by wind or water. About 10% of pollination occurs this way. But most plants (80%) depend on animals for cross pollination. Many flowers produce fragrant smells and sweet nectar to attract bees, hummingbirds, other insects, and even certain kinds of bats. These animals feed on the nectar and pick up pollen on their bodies. Then when they fly to another plant the pollen is transferred to the female parts where fertilization takes place.

Trees

Certain types of trees and shrubs called *conifers* produce pollen and seeds in *cones* instead of flowers. Pines, spruces, firs, and cypresses are examples of conifers. Their leaves look like pointed needles. The female cones that produce the seeds are much larger than the male cones. The conspicuous hard, woody, overlapping *bract scales* of the cones are modified leaves. Lying just above these are thin, flat *seed scales*. Each seed scale produces two *ovules*. When the ovules are fertilized by pollen from the male cones, they develop into seeds.

Tomato Plant Life Cycle

Tomato plants undergo a series of life stages, starting at the seed. The **seed** contains all of its own nutrition for the beginning stages of life. Once it is planted, it grows into a **seedling** and starts using photosynthesis to harness energy from the Sun for life and growth. When the plant is **mature**, it produces **flowers**, which then produce the tomato fruit. The **fruit** contains seeds which then go on to become the next generation for this plant. The tomato plant usually dies after one year, so the next generation comes directly from the seeds of the previous generation.

Lima Bean Life Cycle

Lima beans have a similar life cycle to that of a tomato plant. It begins as a seed, which grows roots underground. As it grows and emerges from the soil, it becomes a seedling, which eventually becomes an adult lima bean plant. Rather than flowering, it produces leaves and pods, which contain several new seeds. These seeds are known as beans, which can be replanted or used for food.

Radish Life Cycle

Radishes also begin as a seed, which germinates into a sprout. The radish grows large leaves above ground while the root underground also grows large and round. When the plant grows into an adult, it flowers, producing more seeds. The large root underground is what most people think of as a radish, and the leaves that grow above ground are known as radish greens. Both the root and the leaves can be eaten.

Frog Life Cycle

Frogs have a very distinct life cycle. Frogs start as eggs, which hatch into tadpoles. Tadpoles live and breathe completely in the water using gills and have no legs or arms. As the tadpole begins to mature, it grows legs and eventually becomes a young frog. The young frogs then emerge from the water and usually live on land and breathe air using lungs. Some frogs are still able to live and breathe underwater throughout their adult life. Frogs then reproduce by laying eggs in the water.

Lady Beetle and Butterfly Life Cycles

Lady beetles, commonly called ladybugs, and butterflies, start their lives as eggs, which then hatch into larvae, which are similar to worms. In butterflies, these larvae are called caterpillars. The larva usually spends its life eating to build up energy for the change to its next stage. The larva eventually turns into a pupa or a chrysalis, which is a far less active stage. In this change, the larva spins a web around itself, becoming a cocoon. In this stage, the body changes form and eventually, an adult butterfly or ladybug emerges. The adult forms of both of these creatures have wings and are then able to fly and eventually produce new eggs.

Cricket Life Cycle

Crickets undergo a similar change to metamorphosing insects. They start as eggs, which hatch into **nymph** crickets. These are the same as the adult form of the insect, only they are smaller and not yet capable of reproducing. As the nymph grows, it eventually becomes an adult and can then mate and lay eggs.

Earth/Space Systems and Earth Resources

Rock Cycle

The **rock cycle** is the process the materials that make up the Earth go through to transition through the three types of rock: igneous, sedimentary, and metamorphic. Rocks, like all matter, cannot be created or destroyed. However, they go through a series of changes and take on different forms through the rock cycle. The rock cycle is a result of plate tectonics and the water cycle. These processes cause rocks and minerals to change so they can exist in different external conditions.

The rock cycle does not perfectly describe the changes that happen to all rocks. However, it describes the typical process that rocks may undergo. In the rock cycle, rocks beneath Earth's surface melt into **magma**. This magma either erupts through volcanoes or remains inside the Earth. The magma cools, forming **igneous rocks**. Igneous rocks on the surface of the Earth are changed by **weathering** and **erosion**, which are processes that can break rocks down and spread their fragments across the Earth's surface. These fragments form layers and can eventually become **sedimentary rocks**. Sedimentary rocks then either transform into **metamorphic rocks**, which will become magma inside the Earth, or directly melt down into magma.

Formation of Sedimentary Rocks

Sedimentary rocks are rocks that are formed from small pieces of minerals and other rocks. These smaller rock and mineral particles are called **clasts**. When rocks that lie on Earth's surface are exposed to the elements, they become **weathered**. This process causes the rocks to break down, which creates clasts. Through **erosion**, these clasts are **deposited** in one location, often under water, and collected together. Clasts are irregular in shape, so as they collect, **pore spaces** form between them. As more and more clasts are deposited, the weight of the added clasts pushes the pieces closer together, reducing the pore space. This process is called **compaction**. However, some pore space will remain after compaction. Water can carry tiny pieces of minerals, some of which may even be dissolved in the water, through the pore space. Those minerals will eventually harden and fill the remaining pore space. This is called **cementation** and is the final step in the formation of sedimentary rock.

Fossil Fuels

Sedimentary rock can contain the remains of living organisms. This happens when clasts are deposited on top of remains or pieces of remains are deposited among the clasts. Over time and with pressure, these remains become substances called **fossil fuels** that can fuel machines. Some examples of fossil fuels are oil and coal.

Example

Which of the following best explains how fossil fuels are formed?

a. Fossils created from the mass extinction of dinosaurs gradually broke down into fuel such as oil.
b. Decaying plants and animals at the bottom of the ocean were compressed into fuel such as oil.
c. The remains of animals encased in ice from the last ice age melted, releasing fuel such as oil.
d. Fossils and bones are processed by humans to create fuels such as oil.

Choice A states that fossil fuels were formed from dinosaur fossils. The production of fossil fuels is unrelated to fossils, despite their name. Fossil fuels are formed from the remains of ancient plants and animals, not dinosaur fossils. Choice A is incorrect.

Choice B proposes that fossil fuels were formed from the remains of plants and animals in the ocean. Over very long periods of time, pressure and heat can turn these remains into fossil fuels. Choice B could be the correct answer.

Choice C states that fossil fuels are the result of animal remains that have become encased in ice during an ice age melting. When this happens, it typically preserves the remains of the animals, and they do not decay or transform into materials such as fossil fuels. Choice C is not the answer.

Choice D has to do with fossils being processed by humans to create fossil fuels. Fossil fuels do not actually come from fossils, and they are created through natural processes over hundreds of millions of years, not by humans. Choice D is incorrect.

Choice B is the correct answer. Fossil fuels are created not from fossils, but from decaying remains of plants and animals in the ocean due to extreme amounts of time, pressure, and heat.

Soils

Soils are formed when rock is broken down into smaller and smaller fragments by physical, chemical, and biological processes. This is called **weathering**. Physical processes include **erosion** and **transportation** by water and wind, freezing and thawing, and **slumping** due to gravity. Chemical changes alter the original substances present in rocks and early-stage soils. Biological processes include **burrowing** by animals like earthworms and rodents and **penetration** by plant roots. As plants and animals die and **decay**, soil becomes rich in dark organic matter called humus.

Properties of Soil

Since soil is a mixture of rock fragments and biological materials, it varies significantly. The **composition** of soil determines whether it can support plant life. Several properties of soil can be used to describe its composition.

- **Texture** refers to the size of the soil particles. Soil texture can be classified as sand, silt, or clay, depending on the size and mixture of the particles.
- **Structure** refers to the density and arrangement of the soil particles. Soil can be compacted, making it dense and rock-like, or it can be loose and easy to work with when planting.
- **Porosity** refers to how well water flows through soil. A higher sand content usually allows water to flow through the soil more easily, whereas clay tends to hold onto water.

- **Chemical properties** of soil cannot be identified without testing the soil to find out which elements are present in it.
- **Color** of soil depends on the types of minerals and organic matter in the soil. For example, redder soil may indicate that there is oxidized (rusted) iron in the soil.

Formation of Deltas, Canyons, and Dunes

A **delta** is landform that is created where a river flows into a larger body of water. The river carries sediment. When it reaches the larger body of water, the water from the river spreads out and the sediment is deposited. Sediment is **deposited** by a river when the movement of the river causes it to move to a different area and it is able to fall through the water and settle. An example is the Mississippi Delta, where the Mississippi River meets the Gulf of Mexico. A **canyon** is a deep ravine between two cliffs. Canyons are usually formed by erosion caused by flowing water over extended periods of time. An example of this is the Grand Canyon in Arizona. A **sand dune** is a mound of sand built by a natural force, such as wind, over time. Forces like wind can create sand dunes by moving grains of sand until they form a mound. Great Sand Dunes National Park, located in Colorado, is home to the tallest sand dunes in the United States.

Example

The Grand Canyon, located in Northern Arizona, is the largest canyon in the United States. It reaches a maximum of 6,000 feet deep. Which of the following factors is most likely to be responsible for the formation of the Grand Canyon?

a. Erosion by water and wind
b. Cracking of the soil resulting from hot and dry weather
c. Accumulation of sediment on the sides of the river
d. Tectonic plate movement

Choice A proposes that the Grand Canyon was formed through erosion and weathering. Over very long periods of time, water and wind can have drastic effects on a landscape and even wear away at rock. Choice A could be correct.

Choice B proposes that the Grand Canyon was formed from soil cracking due to hot weather. While this does happen in certain types of soil, it would not be forceful enough to create such a large canyon. Choice B is not the answer.

Choice C proposes that the sediment carried by the river accumulated on both sides, creating the high walls of the canyon. While this is true for the creation of most riverbanks, this process would not be able to create the enormous heights of the Grand Canyon's walls. Choice C can be eliminated.

Choice D proposes that the Grand Canyon is the result of tectonic plate movement. While tectonic plate movement is likely responsible for the formation of the high ground or plateau that the canyon cuts through, it would not have created the canyon itself. This choice is incorrect.

Choice A proposes the best explanation for the formation of the Grand Canyon, which was formed through the effects of moving water over the course of a very long period of time.

Natural Resources, Renewable Resources, Nonrenewable Resources, and Commodities

Natural resources are things provided by nature that have value to humans, such as minerals, energy, timber, fish, wildlife, and the landscape. **Renewable resources** are those that can be replenished, such as wind, solar radiation, tides, and water (with proper conservation and clean-up). Soil is renewable with proper conservation and management techniques, and timber can be replenished with replanting. Living resources such as fish and wildlife can replenish themselves if they are not overharvested. **Nonrenewable resources** are those that cannot be replenished. These include fossil fuels such as oil and coal, and metal ores. These cannot be replaced or reused once they have been burned, although some of their products can be recycled. **Commodities** are natural resources that have to be extracted and purified rather than created, such as mineral ores.

Natural Resources

The term **natural resources** refers to products and energy that can be harvested from the world.

- **Water** is one of the most abundant resources on the Earth and is necessary for life.
- **Natural gas, coal,** and **oil** exist underground and deep in the ocean and can be used as fuel for machines.
- **Trees** can be harvested for wood and paper and other byproducts that are used in daily life.
- **Metals** can be harvested from the ground and are used in many applications, such as building materials and electronics.
- **Sand** can be used to make glass, soaps, and electronics.
- **Sunlight** and **wind** can be harvested using solar panels and wind turbines to generate electricity.
- **Animal products** are used for food or materials in clothing and some manufacturing processes.

Renewable and Non-renewable Resources

Materials and energy on the Earth are classified as either renewable or non-renewable. The term **renewable resources** refers to resources that will not run out due to overuse, or resources that can be easily reclaimed once used. These include wind, water, and energy from the Sun. Some plants and animals grow and reproduce so fast that they are unlikely to become extinct. **Non-renewable resources** include materials that take a very long time to produce, such as fossil fuels. Once these materials are consumed, they may become very difficult to obtain or impossible to create. Because renewable resources do not run out and non-renewable resources do, environmentalists and scientists are always looking for ways to use renewable resources to supply energy and to reduce consumption of non-renewable resources. Reducing the use of resources is known as **conservation**.

Recycling and Protecting the Environment

When trash is thrown away without being separated for recycling, it usually goes to a landfill where it cannot be recovered and reused. Metal and plastic do not break down readily, and those materials are lost if they are sent to the general trash. Aluminum cans, such as soda cans and food cans, can be recycled and used to make new products down the line. Most paper and plastics can also be recycled. Recycling bins are usually located next to normal trash cans and have the recycle symbol marked on them. This is usually three arrows pointing at each other in a triangle:

Weather and Climate

Weather describes conditions or changes in sunshine, temperature, cloud coverage, precipitation, warm and cold fronts, and more during a specific time in a specific area. These components can change throughout a day, week, or month. Meteorologists track and report what is happening in the atmosphere to help predict how the weather will change in the near future. Weather data is collected and studied over long periods of time, often multiple decades, to evaluate the climate of a particular area. **Climate** refers to patterns in weather conditions, such as temperature, precipitation, and any extremes related to them, in a specific area over a very long period of time. A large factor that influences climate is latitude, which is the distance of an area from the equator.

Clouds

Clouds form when water vapor in the atmosphere cools to the point where it **condenses** and becomes water droplets or small particles of frozen ice crystals. Clouds form when moisture is added to the air through evaporation until the air becomes **saturated** and cannot hold any more water. Then, the water vapor can condense into visible droplets.

Precipitation

Precipitation is water that falls back to the surface of the Earth from the atmosphere. This water may be in the form of **rain**, which is water in liquid form. Raindrops are formed in clouds due to **condensation**. When the drops become too large to remain in the clouds, gravity causes them to fall toward the surface of the Earth. Extremely small raindrops are called **drizzle**. If the temperature of a layer of air through which rain passes is below the freezing point of water, the rain may take the form of **sleet**, which is partially frozen water. Precipitation may also fall in the form of **snow**, which is made of ice crystals. When clumps of snowflakes melt and refreeze, **hail** is formed. Hail may also be formed when liquid water gathers on the surface of a snowflake and then freezes.

Temperature

Temperature describes the amount of **heat** in an area. The Earth's tilt and shape impact the average temperatures in each region. The Sun shines on the areas closest to the equator most directly, causing those areas to have warmer average temperatures. All other areas receive energy from the Sun at different angles, causing lower average temperatures. The farther North or South an area is from the equator, the less directly the Sun will shine on them. Since temperature is an important part of the water cycle, it also has an effect on the humidity, precipitation, and air pressure in an area. Since weather processes are like cycles, these factors can also impact the temperature in an area. Temperature can be predicted based on an area's location, the time of year, the moisture in an area, and the air pressure in an area.

Air Pressure

Pressure describes the amount of force applied to an object over a certain area. For example, a person standing on a bed sinks deeper into the bed than a person laying on it. This occurs because the same amount of force is applied to a smaller area, which means the pressure is higher. Pressure applies to all forces that are applied to other objects, including air. The pressure applied by air in a given place is called **air pressure**, or **barometric pressure**. Because air has mass, and gravity is pulling it toward the Earth, air applies force on everything beneath it. Air pressure changes with **altitude**, which is the height of an object above sea level. The lower the altitude, the higher the air pressure. This is because the air at higher altitudes is weighing on the air underneath it. Similarly, air pressure is lower at higher altitudes.

Impact of Temperature and Humidity on Air Pressure

Humidity and temperature also affect air pressure. **Humidity** refers to how much moisture is in the air. Air that contains more water molecules contains more mass than drier air. Therefore, gravity pulls more on air that has higher humidity, which causes increased air pressure. Similarly, **temperature** can affect the density of air. Gases do not have a constant volume, but the space between the molecules in a gas grows and shrinks with changes in temperature, causing changes in air pressure.

Impact of Air Pressure on Climate and Weather

Weather systems are often described based on air pressure. A **high-pressure system** means that an area has higher air pressure, while a **low-pressure system** has a lower pressure. Pressure will move toward **equilibrium**, or a state of **balance**. Therefore, high- and low-pressure systems naturally interact with each other. Low pressure systems try to **pull air** toward them to reach equilibrium. This causes condensation to form and can eventually produce storms. High pressure systems **push air** away from them and are usually associated with sunny or clear weather.

Weather Fronts

An **air mass** is a region of air that has a consistent temperature and amount of moisture throughout. A **front** is the boundary that usually forms between an air mass and the air surrounding it. This boundary may be sloped or vertical. The shape of the boundary depends on the movement of the air masses that it runs between. There are four types of fronts:

- A **cold front** occurs when a cold air mass moves into an area of warmer air. In this situation, the colder, denser air mass remains near the ground, pushing the warmer mass upward. When the ascending air is unstable, this movement may create severe weather, such as thunderstorms.
- A **warm front** is formed when a warm air mass enters an area of colder air. The warmer air ascends, causing precipitation.

- An **occluded front** occurs when a cold air mass moves past a moving mass of warm air near the surface of the Earth. A cold air mass will usually move more quickly than a warm air mass because it has a higher density. The coldest air remains near the ground, forcing the warmer air to rise.
- A **stationary front** forms when warm and cold air meet and neither air mass can apply enough force to move the other. They remain stationary.

Lightning

Lightning is a huge electric spark that can occur inside a cloud, go from one cloud to another, or go from a cloud to the ground. Turbulent rising air and rising and falling raindrops or ice crystals in a thunderstorm cause differences in electric charge in different parts of the cloud and between the bottom of the cloud and the ground. When the difference in charge is large enough, a lightning bolt will discharge, which neutralizes the difference.

Thunder

As a lightning bolt travels through the air, it pushes the air aside faster than the speed of sound. This produces a shock wave of extremely hot air that creates a loud sonic boom, which we hear as **thunder**. If a person can hear thunder, he or she needs to get indoors quickly as possible, since he or she could be struck by lightning.

We hear thunder later than lightning strikes because sound travels much slower than light. Light travels so fast that it is almost instantaneous from one point to another anywhere on Earth. Sound travels much more slowly—about one mile every five seconds. Light travels one mile in about 5 millionths of a second. Therefore, the approximate distance a person is from a lightning flash can be determined if he or she counts the number of seconds between when the lightning flashed and the thunder it made was heard.

Tornado

A **tornado** is a violent rotating column of air that is in contact with both the ground and a cloud. The column is visible because the extremely low pressure causes water vapor to condense and become visible water droplets. Tornados usually stirs up clouds of dirt and debris when they touch the ground. Tornadoes are the most violent storms on Earth, and the strongest spin at 300 miles per hour.

Hurricane

A **hurricane** is a large tropical storm that forms over the open ocean and produces very strong winds and heavy rains. A **tropical storm** forms when warm water evaporates and the saturated air rises and forms a column of condensed water vapor. As the wind speed increases the pressure falls even more and a hurricane can form. Sinking air in the center of the storm produces an **eye** (see the arrow in the image below) where the weather is quite calm and free of clouds.

Measuring Weather

Weather can be measured using a variety of methods. Rainfall, sunshine, pressure, humidity, temperature, and cloudiness can be measured using tools such as **thermometers**, **barometers**, and **rain gauges**. However, the use of **radar** and **satellite imagery** allows meteorologists to observe at weather across large areas, such as entire continents. This helps meteorologists understand and make predictions about current and developing weather systems. **Infrared** (heat-sensing) **imaging** allows meteorologists to measure the temperature of clouds above ground. Using weather reports gathered from different weather stations spread over an area, meteorologists create **synoptic charts**. The locations and weather reports of several stations are plotted on a chart. Analysis of pressures, rainfall, cloud cover, and other factors reported from each location can reveal basic weather patterns.

Water Cycle

The **water cycle** refers to the circulation of water in the Earth's hydrosphere. The **hydrosphere** exists below the surface, on the surface, and above the surface of the Earth. The water cycle is a process that involves five physical actions.

- **Evaporation** refers to liquid water heating up and changing to into a gas known as water vapor.
- **Transpiration** occurs when water inside plants evaporates directly out of plant leaves.
- **Condensation** refers to the water vapor cooling down and beginning to turn back into a liquid form, which causes clouds to form.
- **Precipitation** refers to rain, snow, hail, or sleet that falls from clouds once water vapor has condensed enough.
- The **storage** stage of the water cycle refers to water being stored in the ground, trees, or bodies of water on the Earth. Water is trapped in vegetation, collected in bodies of water on the Earth's surface, or absorbed into the Earth's surface. **Runoff**, caused by gravity, physically moves water downward into oceans or other water bodies.

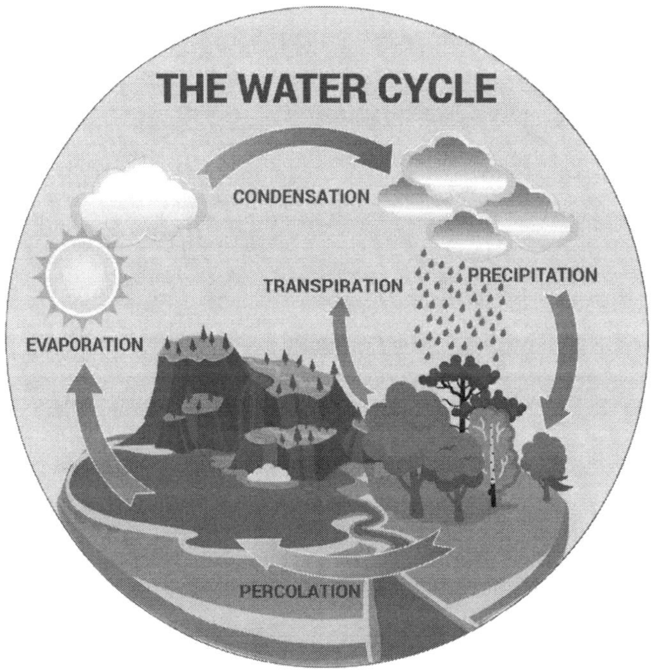

Review Video: Hydrologic Cycle
Visit mometrix.com/academy and enter code: 426578

Roles of the Sun and the Ocean in the Water Cycle

More than 96% of the Earth's water is in the oceans, and the oceans cover more than 70% of the Earth's surface. Ocean water plays a major role in the water cycle. The Sun is also extremely important to the water cycle because it heavily impacts the temperature of the water that sits on Earth's surface. As sunlight hits the surface of the ocean, water heats and begins to **evaporate**. It becomes **water vapor**, which is a gas. The water vapor rises into the atmosphere and **condenses** to form **clouds**. These clouds travel and move across the Earth due to air currents. As more water evaporates and condenses, the clouds become heavy and begin to **precipitate**. This results in **rain**, which collects on the Earth's surface and provides water for living things that need it to survive. These processes also help to regulate the Earth's temperature and prevent large changes in temperature.

Rapid Changes to the Earth's Surface

Normally, the Earth's surface changes with processes such as erosion, weathering, and the warming and cooling of the planet over time. Occasionally, the Earth encounters rapid changes that can be devastating to ecosystems. Examples of natural events that rapidly change the landscape include volcanoes, earthquakes, landslides, hurricanes, tornados, and wildfires.

- **Volcanoes** are mountains or hills that have passages to magma deep beneath the earth. Convection within the Earth's core can cause lava, rock, or gas to travel through these passages to the surface. Volcanoes can be active or dormant.
- **Earthquakes** are strong vibrations in the ground caused by movements in the Earth's crust.
- **Landslides** are sudden shifts of large amounts of land down a hill or mountain as a result of erosion, rain, earthquakes, or volcanic movement.
- **Hurricanes** are large, violent storms with wind and rain arising from warm waters in the ocean.
- **Tornados** are sudden, violent windstorms that can tear apart buildings in a matter of minutes. They often form without warning and usually happen when cold, dry air meets hot, humid air, causing rapid convection.
- **Wildfires** are large destructive fires that spread through woods and brush, destroying habitats. These usually occur in dry environments.

Fossils

A **fossil** is the preserved remains or pieces of an organism that lived thousands of years ago. Examples include bones, shells, and imprints, among others. An **imprint** is created when the organism that died is covered with sediment and acts as a mold within the rock and earth around it. The imprint that remains shows the shape and some more detailed features of the organism. Fossils give scientists useful information about the history of Earth. Different kinds of fossils are found in rocks of different ages, allowing scientists to know the age of a rock or rock layer. Fossils also give scientists information about the organism's behavior and patterns. For example, animal fossils can show scientists how animals traveled, what they ate, and what kinds of habitats they lived in. Both plant and animal fossils can give scientists information about the environment during the time that the organisms were alive. Since the organisms would have needed to live in suitable environments, scientists can use fossils as indicators that the environment had conditions that benefited the fossilized organisms. The features of these fossilized organisms can support scientific conclusions about what the climate was like, whether there were bodies of water nearby, and what other organisms lived in the area.

Plate Tectonics

Main Concepts

Plate tectonics is a geological theory that was developed to explain the process of continental drift. The theoretical separation of the Earth's lithosphere and asthenosphere is based upon the mechanical properties of the materials in the two respective layers and is distinct from the chemical separation of Earth's crust, mantle, and core. According to the theory of plate tectonics, the Earth's lithosphere is divided into **ten major plates**: African, Antarctic, Australian, Eurasian, North American, South American, Pacific, Cocos, Nazca, and Indian; it floats atop the asthenosphere. The plates of the lithosphere abut one another at plate boundaries (divergent, convergent, or transform fault), where the formation of topological features of Earth's surface begins.

Seafloor Spreading

Seafloor spreading was originally put forth as an explanation for the existence of midocean ridges such as the Mid-Atlantic Ridge. These ridges were identified as features of a vast undersea mountain system that spans the globe. Seafloor spreading postulates that the ocean floor expands outward from these ridges. The process occurs when the upper mantle layer of the Earth (the asthenosphere), just beneath the planet's crust, is heated through convection. The heat causes the asthenosphere to become more elastic and less dense. This heated material causes the crust to bow outward and eventually separate. The lighter material then flows out through the resultant rift and hardens, forming new oceanic crust. If a rift opens completely into an ocean, the basin will be flooded with seawater and create a new sea. Often, the process results in failed rifts, rifts that stopped opening before complete separation is achieved.

Continental Drift

Continental drift is a theory that explains the separation and movement of the continents based on shifts in a plastic layer of Earth's interior caused by the planet's rotation (seafloor spreading). Continental drift is part of the larger theory of plate tectonics. In the early twentieth century, many scientists and scholars noted that the edges of certain continents seemed to look like connecting pieces of a puzzle. Due to this observation, as well as the fact that similar geologic features, fossils, fauna, and flora existed on the Atlantic coasts of continents like South America and Africa, these observers theorized the previous existence of a supercontinent (referred to as Pangaea), in which all of the discrete continents identifiable today were joined together.

Theory

This theory of plate tectonics arose from the fusion of **continental drift** (first proposed in 1915 by Alfred Wegener) and **seafloor spreading** (first observed by Icelandic fishermen in the 1800s and later refined by Harry Hess and Robert Dietz in the early 1960s) in the late 1960s and early 1970s. Prior to this time, the generally accepted explanation for continental drift was that the continents were floating on the Earth's oceans. The discovery that mountains have "roots" (proved by George Airy in the early 1950s) did not categorically disprove the concept of floating continents; scientists were still uncertain as to where those mountainous roots were attached. It was not until the identification and study of the Mid-Atlantic Ridge and magnetic striping in the 1960s that plate tectonics became accepted as a scientific theory. Its conception was a landmark event in the field of Earth sciences—it provided an explanation for the empirical observations of continental drift and seafloor spreading.

Tectonic Plate Motion

The two main sources of **tectonic plate motion** are **gravity** and **friction**. The energy driving tectonic plate motion comes from the dissipation of heat from the mantle in the relatively weak asthenosphere. This energy is converted into gravity or friction to incite the motion of plates. Gravity is subdivided by geologists into ridge-push and slab-pull. In the phenomenon of **ridge-push**, the motion of plates is instigated by the energy that causes low-density material from the mantle to rise at an oceanic ridge. This leads to the situation of certain plates at higher elevations; gravity causes material to slide downhill. In **slab-pull**, plate motion is thought to be caused by cold, heavy plates at oceanic trenches sinking back into the mantle, providing fuel for future convection. Friction is subdivided into mantle drag and trench suction. Mantle drag suggests that plates move due to the friction between the lithosphere and the asthenosphere. Trench suction involves a downward frictional pull on oceanic plates in subduction zones due to convection currents.

Review Video: Plate Tectonic Theory
Visit mometrix.com/academy and enter code: 535013

Results of Tectonic Plate Motion

Earthquakes result from the movement of a dozen or so major lithospheric (crustal) plates that float upon Earth's mantle (asthenosphere). These **plates** move about each other, at transform boundaries, in response to complex convection cells set in motion by Earth's interior heat. Two plates move apart from each other at **divergent boundaries**, or spreading centers, and come together at **convergent boundaries**. When thin, denser, iron- and magnesium- rich oceanic crust collides with thicker, lighter, silica-rich continental crust, the former is **subducted** beneath the latter. The subducted material carries scraped-off continental crust and seawater down with it. As this material melts, it rises as a mixture of magma and steam to produce explosive **volcanic mountain ranges** such as those surrounding the Pacific Ocean Basin. A **hot spot** is an area of the Earth's surface that experiences high levels of volcanic activity, created by hypothetical phenomena called mantle plumes. Mantle plumes are exceptionally hot portions of the Earth's mantle that rise (through convection) to the crust. Hot spots are responsible for the formation of volcanic chains such as the Hawaiian Islands. When two continental plates collide at convergent boundaries, the crust buckles and thrusts up massive mountain ranges such as the Alps and Himalayas.

Example

The Andes volcanic mountain range on the western coast of South America is the longest mountain range in the world, consisting of hundreds of volcanoes. Which of the following processes is most likely responsible for the formation of these volcanoes?

a. Convergence of a continental plate and oceanic plate
b. Two tectonic plates moving side-by-side
c. Divergence of two tectonic plates
d. Existence of a hotspot

Choice A describes a continental plate, also referred to as a land plate, colliding with an oceanic plate. When this happens, the oceanic plate slides underneath the continental plate, which is called subduction. This results in long chains of volcanoes, such as the Cascades in the United States, along with earthquakes and sometimes tsunamis. The Andes are a long chain of volcanoes, so Choice A could be correct.

Choice B describes two tectonic plates moving side-by-side. When this happens, it causes earthquakes and rolling hills, not volcanoes. Choice B is not the answer.

Choice C describes the divergence, or separation, of two continental plates. This results in large troughs in the land called rift valleys, not volcanoes. Choice C is not the answer.

Choice D describes the existence of a hotspot, which is a high-temperature spot below Earth's crust that causes magma to rise up through the crust. Hotspots form individual volcanoes or small volcanic groups, such as those in Hawaii and Yellowstone. However, the Andes are a long chain of volcanic mountains, which are not associated with hotspots. Choice D is incorrect.

Choice A is the correct answer. Chains of volcanic mountains like the Andes are formed through subduction when an oceanic plate slides underneath a continental plate.

Oceans

Chemical Composition of Oceans

Salinity is the proportion of salts (inorganic minerals) dissolved in the ocean. These minerals are present as ions within seawater. The proportions of the minerals most commonly found in seawater are as follows: chloride (55%); sodium (31%); sulfate (8%); magnesium (4%); calcium and potassium (1% each); bromide, boric acid, bicarbonate, fluoride, and strontium (less than 1% each). While seawater's salinity may vary, the proportion of salt minerals remains constant. Carbon dioxide, nitrogen, and oxygen are present as dissolved gases in the oceans. The nutrients dissolved in seawater include nitrogen, phosphorus, and silica. The salinity of the oceans is affected by weathering and erosive processes of rocks on land by precipitation (which eventually returns to the seas as runoff), volcanic emissions, and the activity at hydrothermal vents on the deep seafloor. Deep sea vents are created at locations near underwater ridges when mantle material causes the eruption of water containing minerals (such as magnesium) through the lithosphere. The salinity of the ocean affects its electrical conductivity and density.

Review Video: Boric Acid and Borax
Visit mometrix.com/academy and enter code: 954479

Island Arc, Oceanic Trench and Continental Shelf

An **island arc** is one potential result of the convergence of two oceanic plates; it is a type of **archipelago** (chain of islands). When one plate subducts the other, the mantle of the subducting plate is partially melted by the increase in temperature of the subducted plate, which also drives water up into the overriding plate's mantle. As the temperature of the mantle increases and absorbs water, it becomes magma riddled with dissolving gases. This low-density material enters and exits the overriding plate. This process creates **volcanic chains**, such as the Pacific's Ring of Fire, which go through alternate periods of high and low activity. An oceanic trench is a long, deep, narrow depression in the oceanic lithosphere. The **Mariana Trench**, located in the Pacific Ocean, boasts the deepest measured location: 10,911 meters below sea level. A **continental shelf** is the submerged edge of a continent. These shelves are often eliminated when an oceanic plate slips under a continental plate at a converging plate boundary.

A **continental shelf** is a portion of the ocean floor that exists as a platform shallowly submerged off of a beach. This platform has a slight downward grade; its surface is uneven, with a relief (due to glacial activity) up to 100 meters. **Sediment** and **sedimentary rock** make up most continental shelves. The surface of a continental shelf is affected by natural dams, such as tectonic dams (rocks forced upward by tectonic motion) and diapirs (domes of salt that form under sedimentary rocks and push up through the rocks). The presence of these dams creates a depression in which sediment is deposited. A continental shelf experiences an abrupt increase in slope (at an average depth of 133 meters) at the shelf break, the outer boundary of the shelf. The irregular slope that extends downward from the shelf break is called the continental slope; this transition zone connects sea level to the ocean floor (about 3,660 meters below sea level).

Submarine Canyon

A **submarine canyon** is a steep-walled valley in a continental slope. Some believe that these canyons are formed by mass movements of rock along the slope or by turbidity currents (currents initiated by the motion of debris-filled water at the tops of the canyons). It has also been suggested that these canyons were created at a point in time when at least a portion of the relevant area was above sea level; such a point might be the lowering of sea level during the Pleistocene glaciations. These formations occur more often on steeper slopes than on slopes with gentle gradients. Submarine fans (sediment deposits that travel through the canyon and then spill out onto the seafloor), which are similar to land-bound alluvial fans, form the boundary (called the continental rise) between the continental slope and the ocean floor.

Deep Seafloor

The deepest level of the seafloor, which spans from one continental margin to another, is called the **abyss floor**. The flat, relatively featureless aspects of this floor (which is often covered with either basaltic floods or sediment) are called abyssal plains. **Abyssal hills** are fairly gentle rises along the floor of the abyss. The **ocean ridge system**, which is tall enough in some places to break the ocean surface, stands in sharp contrast to the abyssal plains. The ridge system, including the Mid-Atlantic Ridge, includes the median rift, a long, continuous depression that follows the crest. The floor of the deep ocean also exhibits trenches (the deepest part of the ocean) and chains of volcanic islands called island arcs. Basaltic seafloor volcanoes called **seamounts** are reminiscent of intermediate land volcanoes. Seamounts are not tall enough to reach the ocean's surface. Guyots are enormous undersea volcanoes whose apexes appear to have been shorn off, creating a nearly level underwater shelf. Oceanographers have been unable to establish a consensus explanation of this phenomenon.

Coral Reefs

A **coral reef** is an underwater formation composed mainly of coral (small marine creatures similar to anemones) and algae. These often-beautiful undersea features usually form in warm, clear water. They develop their own ecosystems, composed of a wide variety of plant and animal species. Before he became (in)famous for his theory of genetic evolution, Charles Darwin spent time during an ocean voyage studying (among other things) coral reefs. He identified three major varieties of reef and postulated a successive relationship between them: A **fringing reef** (which extends out from and grows along a shoreline) begins to grow around a sinking volcanic island. The fringing reef becomes a **barrier reef** surrounding the still-sinking island when the portion of the reef closest to the shore cannot keep up with the biotic (live) growth of the outer potion; the inner portion becomes a lagoon. This process continues, and the reef continues to spread around the island in a circuit. When the island sinks completely, the coral formation has become an **atoll**.

SEDIMENTS ON THE OCEAN FLOOR

Much of the abyss floor is covered with sediment. The amount of sediment in any location on the ocean floor is related to the age of that portion of the oceanic lithosphere. **Ooze** is a type of sediment that accumulates over time from fragments of deceased surface organisms. The type of ooze deposited varies with depth and chemical conditions. **Calcareous ooze** is ooze made up mostly of the carbonate remains of organisms such as shells. This type of material dissolves at depths below four kilometers. **Siliceous ooze** is that composed of the remains of silica-cell–producing organisms. This type of ooze also dissolves under certain conditions; however, it is more resistant to dissolution than carbonate material. Sediment at the deepest levels of the ocean consists largely of (inorganic) clay, with the occasional appearance of fossils like sharks' teeth. Oceanic sediments provide large amounts of information about the Earth's prehistory, due to its sheer volume and the fact that fossils survive better undersea (without oxygen) than fossils on land.

SEA BREEZES

The daily influxes in the strength and direction of wind at ocean coasts are collectively referred to as sea breezes. These phenomena contribute to the climate-mediating function of oceanic bodies. When solar energy hits the coastal surface during a sunny day, the ground warms quickly. Air above this heated land rises through convection, creating a low-pressure zone. Cooler air then flows into the region vacated by the risen air, creating a cool onshore breeze (unlike meteorologists, oceanographers name winds for the direction they are headed) by the middle of the day. Conversely, when the shore cools after the Sun goes down, the air above it cools and sinks. The direction of wind flow is then reversed, creating an offshore breeze in the evening.

SEAWATER MOTION AND ATMOSPHERIC WIND PATTERNS

We can envision the ocean as a series of **layers of water**, loosely connected to one another by friction. Winds over the ocean move the surface layer of water, which exerts a force on the layer below it, which affects the layer below it, and so on. However, the energy generated by the wind decreases with ocean depth, so that after a certain point (usually a depth of 100–200 meters), the wind has little if any effect on the water. Due to the influence of the Coriolis force, surface currents do not move in the same direction as the winds above it; ocean waves seem to move to the right of the wind in the area. This can be explained in terms of **Ekman transport**. This theory states that as each layer of oceanic water is touched by wind, it also experiences the effects of the Coriolis force, which causes it to move 90 degrees to the right (or left, in the Southern Hemisphere) and spiral downward (an Ekman spiral).

SEAWATER CIRCULATION

At the surface, **atmospheric winds** and the **Coriolis force** are the most influential factors in seawater circulation. Surface waters move in opposite directions on either side of the equator: Northern Hemisphere waters move eastward, and waters in the Southern Hemisphere tend to move westward. In a manner similar to that in which air circulates in the atmosphere, warm surface waters travel from regions near the equator toward the cooler poles, redistributing heat within the oceans. These wide, circular patterns are called gyres. When warm surface water reaches the higher latitudes, it is cooled by interaction with the ambient waters (and ice) and the air in the region. With the resulting increase in density, the water moves downward and cycles back toward the equator. Thus, gravity is the force that most affects deep-water circulation. Some of the water in this "oceanic conveyor belt" then rises back to the surface where it is warmed again and reenters the cycle. Other portions of water remain in the depths of the ocean.

Waves

There are two main processes in operation in the creation of wave motion: disturbing forces (such as wind, changes in atmospheric pressure, the mixing of water of different densities, and earthquakes) and restoring forces (the water's surface tension and gravity). In the creation of a wave, a **disturbing force** drives water into a pile, or crest. The water that makes up this crest comes from an adjacent section of water; therefore, the formation of the crest produces a depression, or trough, on a nearby water surface. Gravity, a **restoring force**, attempts to restore equilibrium by forcing the crest back down to sea level. Due to inertia, however, the gravitational force presses water down below that level, creating another trough. The water displaced by this motion forms another crest, which produces another trough, and the up and down motions continue. When the crest of a wave is higher than its associated trough, the wave moves forward. It is not the water that moves during wave motion, but the wave.

Wave length is defined as the horizontal distance between two equivalent wave phases (that is, the distance separating two sequential crests or troughs). The **velocity** of a wave is the distance it moves during a specified period of time. The **period** of a wave is the length of time required for two equivalent wave phases to pass a certain point. The number of periods that occur in a specified length of time is referred to as the **frequency** of the wave. These measurements vary greatly among waves. One of the largest such occurrences measured a wave length of 792 meters, a velocity of 126 kilometers per hour, and a period of 22.5 seconds.

Wave Erosion

Most **coastal erosion** takes place when waves crash into a shore zone during a storm. It is at this time that the height of a wave (the distance between its upper effective limit and its lower effective limit, or wave base) is the greatest. The upper and lower limits of a wave work on a coastline much like a saw moving horizontally. Like river and stream erosion, most wave erosion is accomplished through abrasion caused by eroded materials (sand and gravel, for instance) carried in the water. Fluctuations in the tide cause coastal erosion indirectly—during low tide, large portions of the shore are exposed and vulnerable to other erosive agents, such as wind and rain. The energy of waves and the strength of the material of which a coastline is composed are the main factors in the rate of wave erosion.

Landforms Created

The **sea cliff** is a landform that is often associated with coastal areas. Particularly large, high-energy waves attack coastlines during storms and abrade portions of a rock face, forming a usually steep cliff. The intensity of this effect is related to the durability of the material being eroded and the slope of the land surface. Wave erosion also produces wave-cut platforms, or a truncated sea cliff with a leveled surface. Sometimes these areas remain bare except for the irregularities created by the erosive process; these remnants of cliffs are called stacks. Beaches are created when the portion of the platform closest to the sea is covered with sand or other grainy materials. These formations are constantly being eroded away and reformed. A marine-built terrace (a deposit of material transported by waves) is often formed inland from a beach.

Oceanic Tides

Oceanic tides are the daily oscillations in sea level along coastlines. The number of times sea level changes and the differences in sea level vary with location, but all parts of the ocean experience this phenomenon. Tides occur due to the gravitational force and centripetal acceleration. Both the Sun and the Moon exert gravitational pulls on the Earth; however, the attraction of the Moon is more effective in tide formation, due to the fact that it is closer to our planet (because gravitational attraction varies with distance, among other things). As the Earth rotates, different areas of the oceans are closest to the Moon. Since water is not firmly affixed to the surface, it is disturbed by the Moon's gravitational pull, which creates an exceptionally long wave. Centripetal acceleration is the force that keeps the Earth and Moon in orbit around the Sun; this force is experienced equally at all points on the Earth's surface. The combined influence of these two forces produces the rhythmic pattern of oceanic tides.

Review Video: Spring and Neap Tides
Visit mometrix.com/academy and enter code: 825792

Coastal Upwelling

Coastal upwelling occurs due to the combined effects of wind and Ekman transport. Wind often blows parallel to a coastline; Ekman transport causes warm surface water along the shore to swing outward, away from land. Coastal upwelling happens when cooler water swells up from higher depths to replace the water moving away from the coast. The water that rises during coastal upwelling contains high levels of nutrients. Plants near the surface absorb these nutrients and consequently experience population growth. Fish and other marine organisms then come to the surface to feed on the plentiful vegetation. Thus, areas of coastal upwelling are some of the most fertile in the ocean.

Effects of Ozone Depletion on Marine Life

Ozone depletion affects marine life due to the harm caused by ultraviolet radiation. In the oceans, this electromagnetic energy can penetrate depths to 20 meters. This radiation can cause cell damage in marine species. Small organisms, such as plankton (plant and animal species), are especially susceptible to damage. When these organisms, which exist on the lowest tier of the food chain, are killed or mutated, effects radiate upward on the food chain. This process has the potential to decrease the global food supply. Ultraviolet radiation also affects fish, amphibians, shrimp, and other organisms in their developmental stages, decreasing their future reproductive capabilities. Again, this damage can move up the food chain and impact entire ecosystems.

Effects of Pollution on Oceans

Sewage, oil, garbage, and industrial waste are commonly disposed of in Earth's oceans. Also, the use of pesticides in agriculture can indirectly contribute to pollution—these chemicals are picked up by rainwater moving back to the oceans. **Ocean pollution** has many harmful effects, both within oceanic ecosystems and across the rest of the globe. For example, pesticides deposited in the ocean contaminate the normally clean water in which algae thrive. When the algae die, a number of the fish that depend on algae (or other underwater plant life) for sustenance will die as well. This can affect local economies (not to mention food supplies) that depend on fishing. Also, when toxic waste is dumped into an ocean, it may be consumed by organisms low on the food chain. If it does not kill these organisms, they can be eaten by larger predators. Pollutants can continue to move up the food chain and eventually be consumed by humans. Lead, for instance, is commonly found in seafood eaten by humans.

Solar System

The **solar system** consists of the **Sun** and eight major **planets**. In order from the Sun, the planets are Mercury, Venus, Earth, Mars, Jupiter, Saturn, Uranus, and Neptune. Pluto is no longer considered to be a major planet. Pluto and five other similar-sized objects are considered **minor planets**. Six of the major planets have one or more moons. The solar system also contains countless meteoroids, asteroids, and comets.

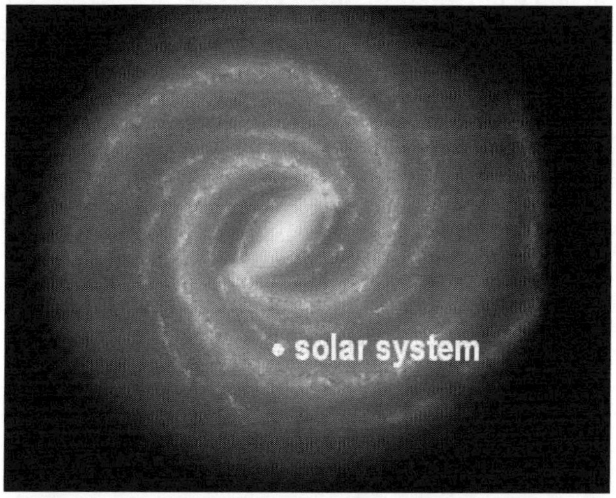

Characteristics of the Sun, the Moon, and Earth

The Sun, Earth, and Moon are all part of the solar system. They are all spherical in shape and rotate on their own axes. However, they each have different physical characteristics and patterns.

- **The Sun:** The Sun is much larger than Earth and is the center of the solar system. The planets orbit around it. It is made up of gases. The most abundant gas is hydrogen, but it also contains helium and a small amount of metal. The temperature of the Sun's surface is nearly 10,000 degrees Fahrenheit.
- **Earth:** Earth orbits around the Sun over the course of a year, rotating on its own axis at a constant tilt. The tilt causes the change in seasons. Earth makes one complete rotation on its axis in 24 hours. The distance between the Sun and Earth allows Earth to have a temperature that sustains life, unlike planets that are closer to, or farther away from, the Sun. Earth is made up of solids, liquids, and gases. Oxygen and iron make up the majority of Earth's mass.
- **The Moon:** The Moon also rotates on its own axis, but it orbits around Earth and is much smaller than Earth. The Moon is made up of solid rock. The surface of the Moon contains many depressions called craters.

Earth's Rotation

The **Earth rotates** west to east about its axis, an imaginary straight line that runs nearly vertically through the center of the planet. This rotation (which takes 23 hours, 56 minutes, and 5 seconds) places each section of the Earth's surface in a position facing the Sun for a period of time. This creates the alternating periods of light and darkness we experience as **day and night**. This rotation constitutes a **sidereal day**. It is measured as the amount of time required for a reference star to cross the **meridian**, an imaginary north-south line above an observer. Each star crosses the meridian once every **sidereal** day. Since the speed at which Earth rotates is not exactly constant, we use the average **solar day** (a 24-hour period) in timekeeping rather than the slightly variable sidereal day.

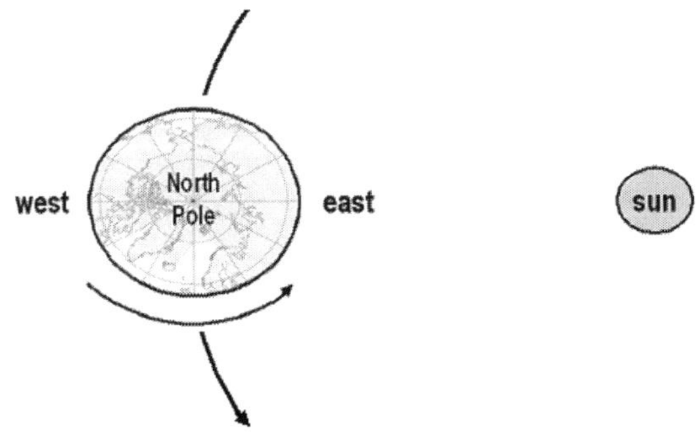

Earth's Revolution Around the Sun

Like all celestial objects in our solar system, **Earth** revolves around the **Sun**. This process takes approximately 365 1/4 days. This period of time is equal to a **calendar year**. The path of the orbit of Earth around the Sun is not circular, but it is **elliptical**, or shaped like an oval. Therefore, the distance between the Earth and the Sun at points on either extreme of this counterclockwise orbit are not equal. In other words, the distance between Earth and the Sun varies over the course of a year. At **perihelion**, the Earth's minimum distance from the Sun, Earth is 147 million kilometers from the Sun. At **aphelion**, the Earth's maximum distance from the Sun, Earth is 152 million kilometers from the Sun. This movement of the Earth is responsible for the apparent annual motions of the Sun (in a path referred to as the ecliptic) and other celestial objects visible from Earth's surface.

The Appearance of the Sun

The Sun is the largest object in the solar system, and it is about 109 times larger than Earth. The Earth is constantly in **orbit** around the Sun. While the Earth is in orbit, it is also **rotating** on its own **axis**. Because the Earth is always moving in this way, it appears that the Sun is moving across the sky and changing location over the course of the day. However, the Sun is not actually moving across the sky. The region of the Earth that is facing the Sun experiences daytime, while the rest of the Earth experiences nighttime. In addition, the path that the Earth travels as it orbits the Sun is not a perfect circle. It is more **elongated**, like an oval. This means that the distance from the Sun to Earth varies throughout the year. The changing distance affects how bright the Sun looks from Earth.

The Sun and Seasons

The orbital path the Earth follows around the Sun helps create the **seasons**: winter, spring, summer, and fall. Earth is closest to the Sun in January and farthest in July. However, there is also a tilt to the Earth's axis. During July, the Northern Hemisphere is tilted towards the Sun. This causes the light and heat the Sun gives off to reach the Northern Hemisphere at a more direct angle. This is why July has hotter temperatures than January in the Northern Hemisphere, even though Earth is farthest away from the Sun during that time. These conditions are the opposite in the Southern Hemisphere, as the region's hottest temperatures occur in January. The Earth's tilt and orbit also affect the number of daylight hours a region experiences at different times, as people in the Northern Hemisphere experience longer days and summertime in July, and people in the Southern Hemisphere experience shorter days and winter in July.

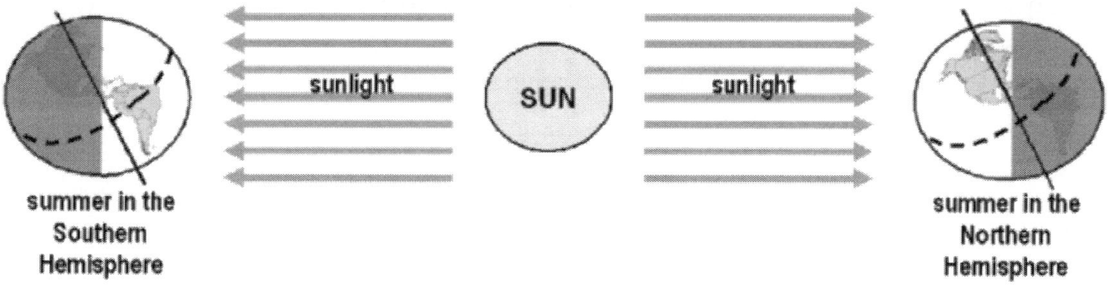

Phases of the Moon

The light of the Moon is a reflection of sunlight that reaches the Moon. The Moon is always orbiting Earth, and as Earth and the Moon both move throughout their orbits, sunlight reaches different areas of the Moon. A **moon phase** describes the appearance of the Moon. Moon phases depend on the position of the Moon, Earth, and Sun. A **new moon** occurs when the Moon is between the Sun and Earth. During this phase, the Moon is not visible because the Sun is illuminating the side of the Moon that is facing away from Earth. When any sunlight reaches the side of the Moon that is facing Earth, the appearance of the Moon will depend on its phase. For example, following a new moon, the right edge of the visible side of the Moon will look like a crescent. This phase is called a **waxing crescent**. During the next phase, the **first quarter**, the right half of the visible side of the moon will be lit up in a half-circle shape. Following that phase, the moon will then appear to be in an oval shape during its **waxing gibbous** phase. After that phase is the **full moon**, where the moon appears as a bright, white circle. Then, the Moon returns to a new moon as it goes through these phases in the opposite order and direction. These phases are known as the **waning gibbous**, **third quarter**, and **waning crescent** phases.

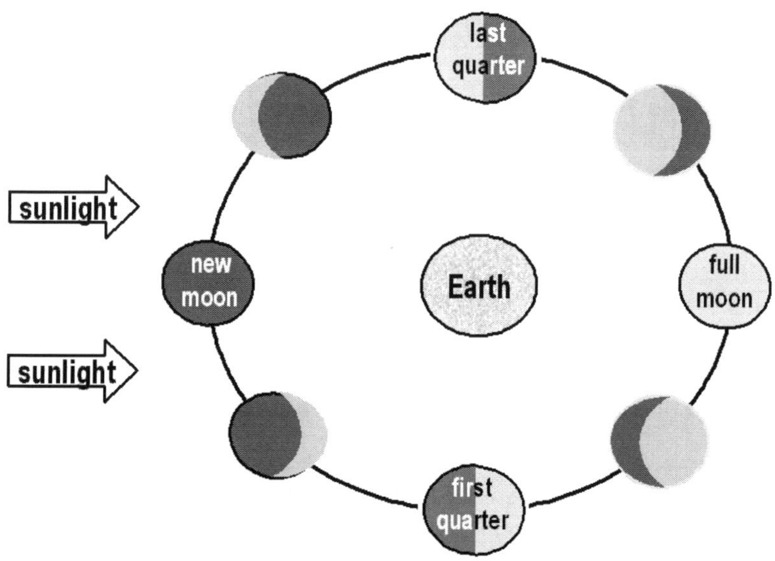

ECLIPSE

An **eclipse** occurs when either the Moon or Earth blocks the light from the Sun. Types of eclipses include solar eclipses and lunar eclipses. A **solar eclipse** is when the Moon is directly in between the Sun and Earth in a straight or nearly straight line. This is very similar to a new moon, but what makes it different is that the Moon is at a specific location between Earth and the Sun. In addition, instead of the Moon appearing to be gone, the moon is outlined with sunlight. A **lunar eclipse** is when Earth is directly in between the Sun and the Moon, in a straight or nearly straight line. Just as a solar eclipse is similar to a new moon, a lunar eclipse is similar to a full moon and occurs when the Moon is at a specific location on the opposite side of Earth as the Sun. During a lunar eclipse, the Moon will still appear as a full circle, but it will appear dim and red instead of bright and white.

Shadows

The Sun provides light to Earth. As the day goes on, the Earth rotates, and the Sun appears to move across the sky. Earth's rotation causes the light from the Sun to reach different parts of Earth at different angles. As the Sun becomes first visible in the morning and appears to rise, it creates long shadows because of where its light hits objects. As the middle of the day approaches, the Sun can be seen high in the sky, and fewer **shadows** are created because the light hits objects from almost directly above. Shadows are also shorter during this time of day. When the Sun begins to set, the shadows will again be longer, but they are cast in the opposite direction as they were at sunrise.

The appearance of shadows also changes due to the seasons. Shadows are smaller in the summer than in the winter. This is due to the tilt of Earth's axis. In the summer in the northern hemisphere, the northern half of Earth is tilted toward the Sun, so the sunlight is more direct and therefore creates a shorter shadow. In the winter in the northern hemisphere, the northern half of Earth is tilted away from the Sun, so the sunlight is not as direct, therefore creating longer shadows.

EXAMPLE

A group of students traces their shadows at 8:00 am in the morning and again at 3:00 pm in the afternoon. They observe that the size and position of their shadow have changed. Which of these explanations for their findings is correct?

a. The Sun has moved in its orbit around the Earth.
b. The students became shorter over the course of the day due to gravity.
c. The Sun's brightness fluctuates over time.
d. The rotation of the Earth has changed the angle of the Sun's rays.

Choice A states that the Sun orbits around the Earth, causing the size and position of the students' shadows to change. The Earth orbits around the Sun, and the apparent movement of the Sun in the sky is a result of Earth's 24-hour rotation. Choice A is incorrect because the Sun does not orbit around the Earth.

Choice B states that the students' shadows become shorter because their bodies have shrunk over the course of the day due to gravity. While gravity has a minute effect on your height as the day goes on, it does not explain the drastic changes to the size of shadows that happen throughout the day. Choice B is incorrect.

Choice C states that the Sun's brightness fluctuates over time, causing the shadows on Earth to change shape. This would not cause the size and position of their shadows to change within a short period of time. Choice C is not the answer.

Choice D has to do with the rotation of the Earth on its axis, which means the angle of the Sun's rays hitting a certain location changes over the course of the day. When the Sun is at its highest point in the sky, that means that this part of the Earth is directly facing the Sun. As the Earth rotates away from the Sun, the Sun begins to set in the sky and the angle of its rays becomes less direct. This change in angle could explain the change in the students' shadows.

Choice D correctly describes the effects of the Earth's rotation on the angle of the Sun's rays, resulting in the Sun's apparent movement across the sky and the changes in shadows over the course of the day.

Practice Test #1

1. In a science experiment, a student is asked to identify the freezing and melting points of water. Which piece of laboratory equipment should be used?

a. Balance
b. Thermometer
c. Ruler
d. Graduated cylinder

2. A science class was asked to illustrate the life cycle of a common insect. One student chose the house fly. The life cycle of a house fly is pictured below.

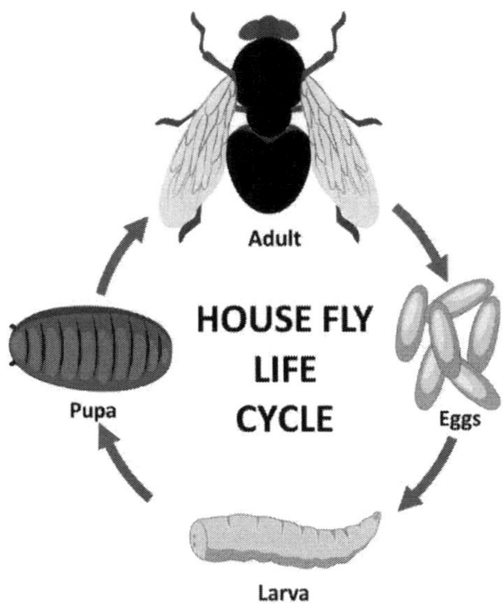

The life cycle of the house fly is most similar to which other life cycle?

a. Butterfly
b. Frog
c. Bird
d. Bat

3. The United States is divided into four time zones. The east coast is three hours ahead of the west coast. This means that when it is 6:00pm in New York, on the east coast, it is 3:00pm in California on the west coast. Based on the information given, which direction does Earth rotate on its axis?

a. Clockwise as viewed from above the north pole
b. Counterclockwise as viewed from above the north pole
c. Toward the Sun
d. The direction varies depending on the season

4. A scientist found this rock full of fossilized teeth.

What can the scientist reasonably conclude about the animal these teeth came from? What evidence supports this conclusion? Write your response to these questions on the lines below.

5. How does the tilt of Earth's axis cause seasons?

a. A hemisphere experiences winter when that half of Earth is tilted away from the Sun. It experiences summer when that half of Earth is tilted toward the Sun
b. A hemisphere experiences spring when that half of Earth is tilted away from the Sun. It experiences autumn when that half of Earth is tilted toward the Sun
c. A hemisphere experiences summer when that half of Earth is tilted away from the Sun. It experiences winter when that half of Earth is tilted toward the Sun
d. A hemisphere experiences autumn when that half of Earth is tilted away from the Sun. It experiences spring when that half of Earth is tilted toward the Sun

6. Students are planning a science experiment to answer a question asked by their teacher. They will release a small toy car down an inclined plane. The students complete their first trial and record how long it takes the toy car to reach the bottom of the inclined plane. For the second trial, the students raise the plane by one level to increase its incline and then record how long it takes the car to reach the bottom. The experiment is done two more times, raising the inclined plane one level higher each time.

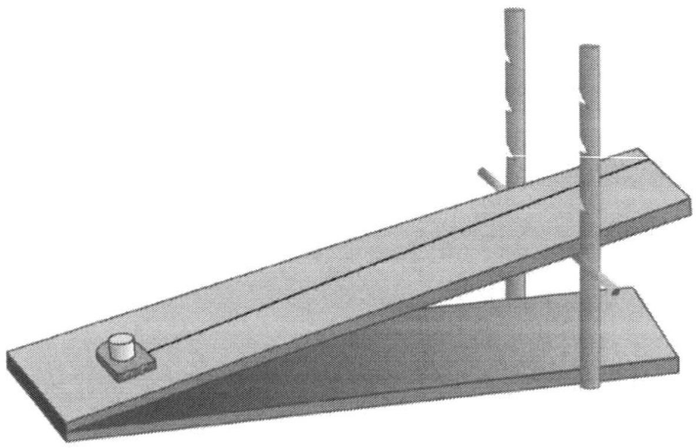

What question are the students most likely trying to answer with this experiment?

a. How can the speed of the car be increased without applying another force?
b. Does the weight of the car affect how fast it can travel?
c. Do different sized cars take different amounts of time to move?
d. How much force does it take to make a toy car move?

7. The poisonous dart frog can be bright blue, bright yellow, and even bright orange. Its coloring may be beautiful, but it is used to scare off predators such as snakes. Dart frogs live near moss close to rivers and streams, but sometimes they can be found in tall trees. They like a warm, yet moist environment.

Ecosystem	Temperature Range	Land Features	Plant Life
Q	Very cold winters with lows well below 0 degrees Fahrenheit and mild summers with highs only around 50 degrees Fahrenheit	Glaciers, ice-capped mountains	Spruce trees, sedum
R	Hot summers and warm winters	Sandy dunes	Cactus, agave, brittlebush
S	Mild winters with average temperatures around 45 degrees Fahrenheit and hot summers with highs well above 100 degrees Fahrenheit	Plains and rolling hills	Oak trees, flowers that like full sun
T	Warm and rainy most of the year	Streams and rivers with lush land	Lots of tall trees and vegetation grow at different levels

Which ecosystem described above would be the best for a poison dart frog?

a. Ecosystem Q
b. Ecosystem R
c. Ecosystem S
d. Ecosystem T

8. How are igneous rocks formed?

a. Years of sediment are laid down on top of each other and forced together
b. Acid rain caused by pollution creates holes in metamorphic rocks
c. Dust and pebbles are pressed together underground from Earth's heat and pressure
d. Magma from a volcanic eruption cools and hardens

9. Algae are part of many food chains. How do algae produce energy?

algae fish seal shark

a. They hunt fish for food
b. Algae undergo photosynthesis
c. They decompose dead or dying organisms
d. They consume fossil fuel from beneath Earth's surface

10. In which circuit will the propeller spin when the switch is closed?

a.

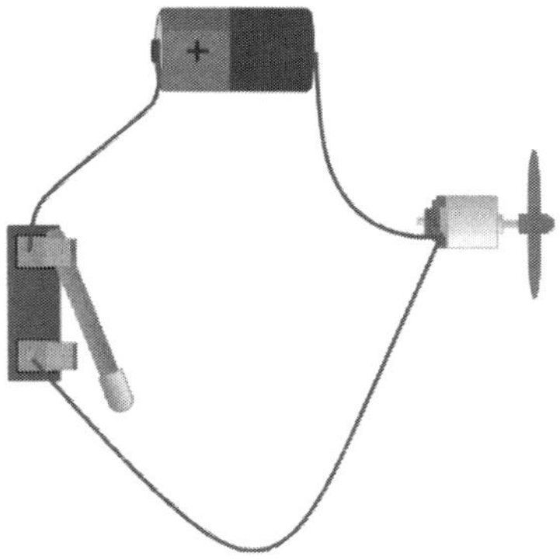

c.

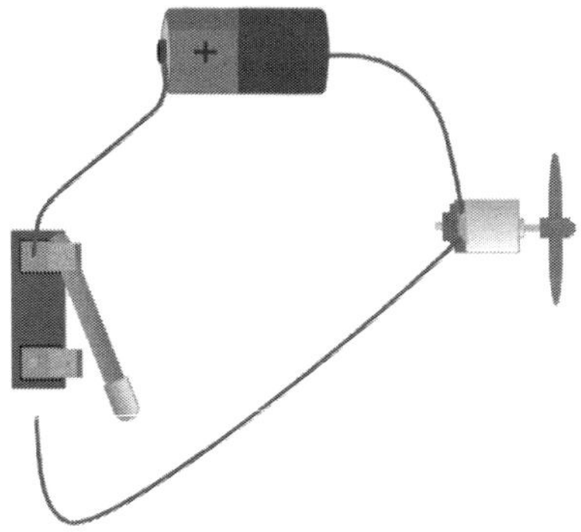

b.

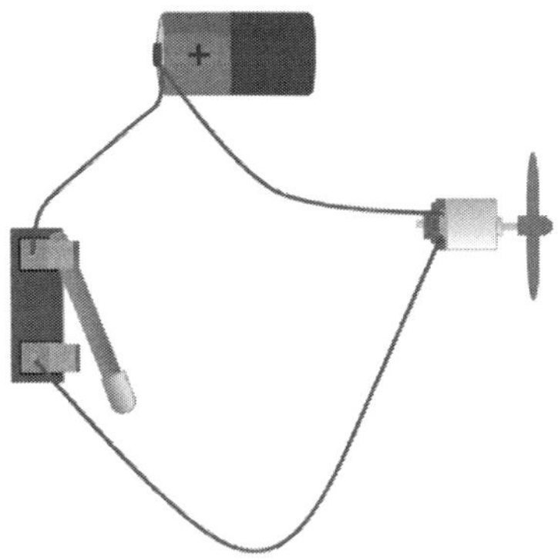

d.

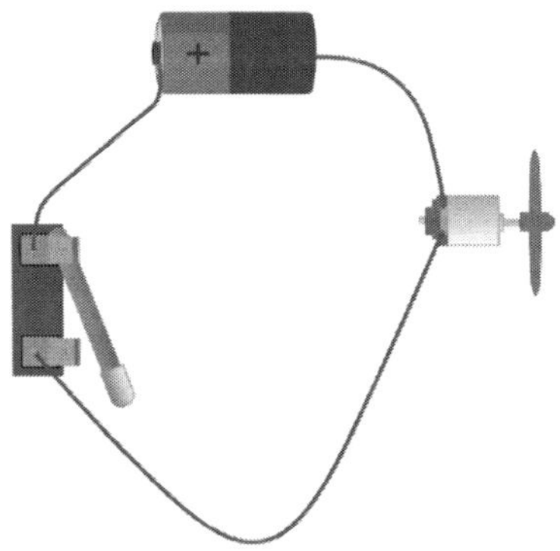

11. Which of the following is the most likely food source for a carnivore?

a. A plant
b. The Sun
c. An animal
d. A dead tree

12. Why are plants considered producers?

a. They make their own food
b. They soak up water from the ground
c. Their leaves can be large or small
d. They produce pollen, which is a food source for many insects and birds

13. How should students determine the volume of fluid in a graduated cylinder?

a. Read the volume at the bottom of the meniscus
b. Read the volume at the highest point the fluid reaches
c. Read the volume at the center, between the meniscus and the highest point of the fluid
d. Read the volume at the highest point then round to the nearest tens

14. Which of the following correctly lists the planets from closest to farthest from the Sun?

a. Jupiter, Venus, Earth, Uranus
b. Saturn, Jupiter, Uranus, Neptune
c. Mars, Earth, Venus, Mercury
d. Venus, Mars, Saturn, Neptune

15. After studying the water cycle, students were asked to fill in the missing parts of the table below.

Water Cycle Term	Explanation
___1___	The process in which liquid is heated by the Sun and turns into vapor (a gas)
Condensation	The process in which the vapor in the air is ___2___ and turns into tiny droplets of water
___3___	The process in which water falls to the surface of the Earth either as a liquid or solid
Collection	The process in which fallen precipitation flows ___4___ and collects in rivers, lakes, and oceans

Which shows the correct terms for each missing part?

a. 1. Precipitation
   2. Cooled
   3. Evaporation
   4. Uphill

b. 1. Evaporation
   2. Heated
   3. Cementation
   4. Downhill

c. 1. Evaporation
   2. Cooled
   3. Precipitation
   4. Uphill

d. 1. Evaporation
   2. Cooled
   3. Precipitation
   4. downhill

16. A ball is thrown and the distance it traveled is measured. Which data set corresponds to the graph below?

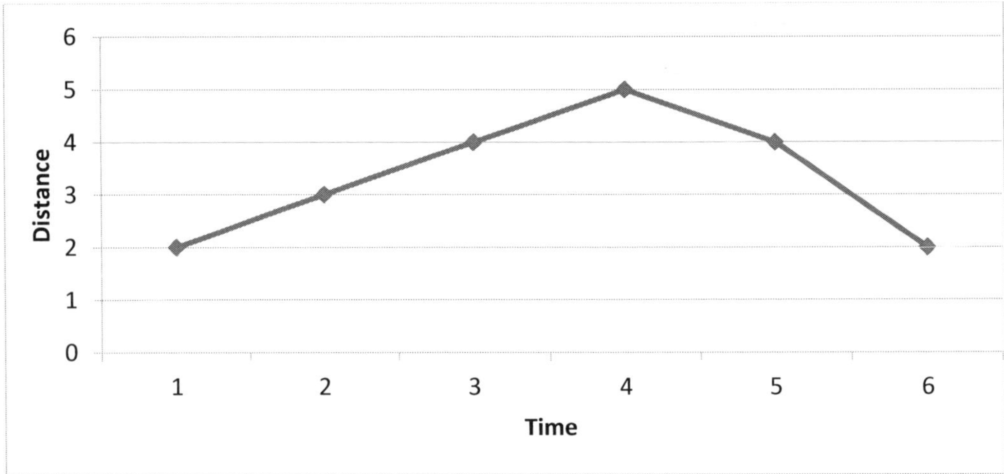

a.

Time	Distance
1	2
2	4
3	6
4	8
5	9

b.

Time	Distance
1	2
3	4
5	6
7	8
9	9

c.

Time	Distance
2	0
4	1
6	3
8	5
9	7

d.

Time	Distance
1	2
2	3
3	4
4	5
5	4
6	2

17. Based on the weather forecast, what can a farmer expect to happen to his bird bath overnight?

10pm	11pm	12am	1am	2am	3am	4am	5am
34°F	33°F	28°F	26°F	24°F	22°F	22°F	21°F

a. Freeze
b. Evaporate
c. Melt
d. Overflow

18. The first microscope was developed by Hans and Zacharias Janssen and allowed scientists to

a. See organisms too small to be seen with the naked eye
b. Count the number of organisms in the universe
c. Determine the causes of all human illness
d. Study why some females are taller than others

19. Electrical cords, such as the cords of a television or the cord on a toaster, are covered with rubber or plastic. What would you expect to happen if a portion of the rubber covering were missing when you tried to plug the cord in?

a. The appliance would not work after it was plugged in
b. The person plugging the appliance in would feel an electric shock
c. The appliance being plugged in would work, but all other appliances in the home would stop working
d. The cover does not matter, all appliances would work normally

20. A student is writing a science fiction story where the Earth suddenly starts rotating more quickly on its axis. This would result in what type of change for the Earth?

a. The days would be shorter.
b. The seasons would happen in a different order.
c. The years would be shorter.
d. There would be no nighttime, only daytime.

21. The students worked in groups to create circuits. One group created the circuit pictured below.

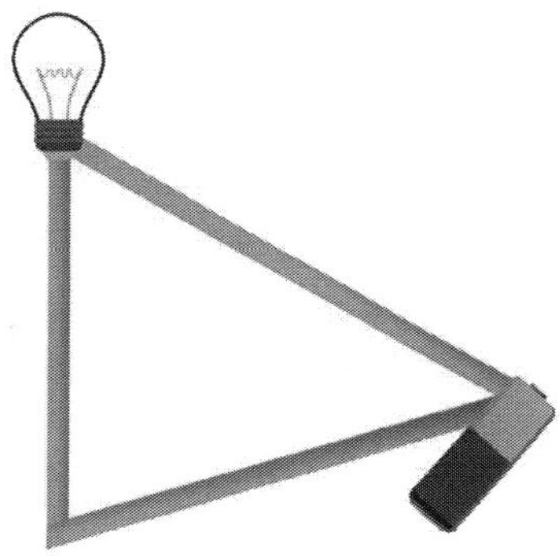

Which conclusion about the circuit is correct?

a. The light bulb will light up because both sides of the light bulb are connected by conducting wires.
b. The light bulb will not light up because the electric current is moving in a diagonal direction.
c. The light bulb will light up because it is a closed circuit.
d. The light bulb will not light up because only one side of the power source is connected by a conducting wire.

22. Which statement accurately describes a state of matter?

a. Solids take the shape of their container
b. Gases maintain a fixed shape
c. Liquids will expand to fill the volume of a container
d. Gases expand to fill an entire space

23. The science teacher wanted to show her students a video. She turned on the projector and pointed it at the screen at the front of the room. One student accidentally stood up directly in front of the projector. The other students could no longer see the light from the projector on the screen. Why could the students not see the light from the projector anymore?

a. Light moves in a straight line and cannot move around another object.
b. Light always scatters when it contacts another solid object.
c. The light was reflected off the student's shirt and was now on the opposite wall.
d. Light moves in an upward direction when another object gets in its way.

24. A carpenter uses all of the tools below on a regular basis.

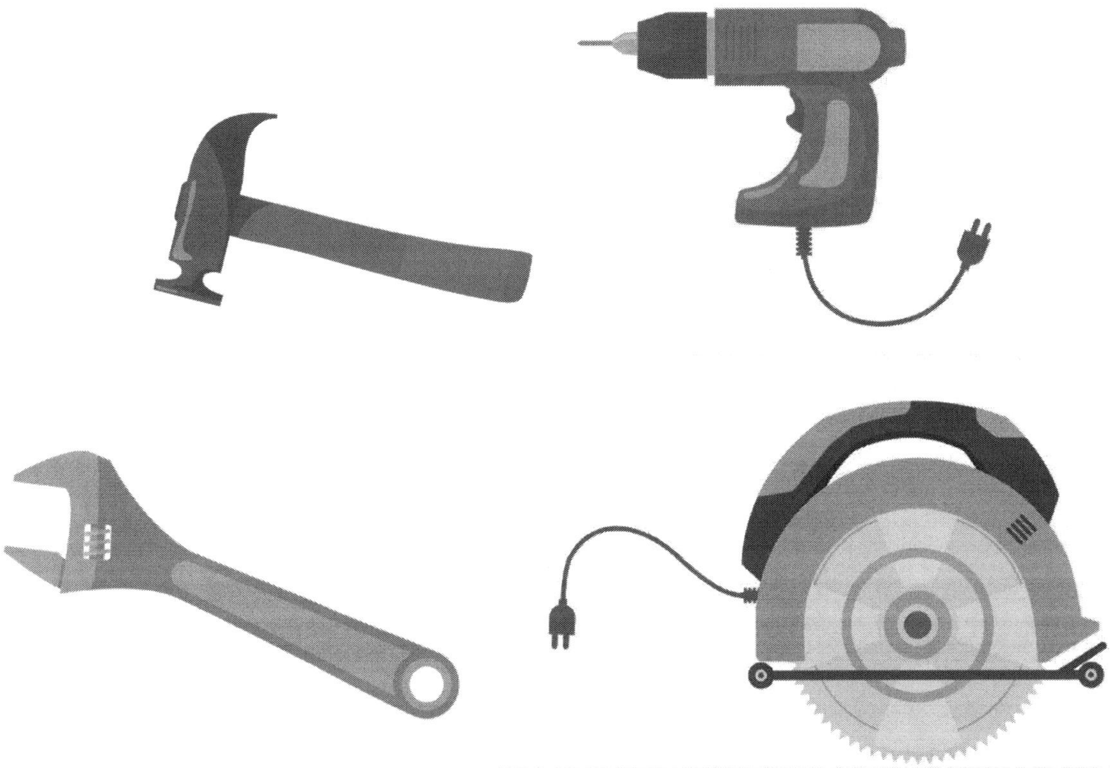

Which type of energy is needed to use all four of these tools?

a. Electrical
b. Thermal
c. Light
d. Mechanical

25. A man accidentally drops his wallet in a swimming pool. He can see his wallet at the bottom of the pool. He jumps in to retrieve it, but the wallet is not where it appeared to be. What is the reason for the optical illusion?

a. The reflection of sunlight off of the water disrupted his view
b. Light is refracted as it enters the water changing the wallet's apparent location
c. The current at the bottom of the pool caused the wallet to move
d. The heat from the Sun has impaired the man's vision

26. A student used dirt, small rocks, and water to create a model of a mountain. She placed the model into an aluminum pan. Next, she poured water into the pan, just until it covered the bottom of the pan. Finally, she moved the pan slowly back and forth. She noticed that after a while, pieces of the mountain started to break away where the water hit the mountain. What science concept is the student exploring?

a. Erosion
b. Melting
c. Decaying
d. Cementation

27. The observations outlined below were made about a pot of water over a campfire. What is the most likely explanation for these observations?

The pot was half full of water
The pot hung over the campfire for 15 minutes
The water began boiling after 7 minutes
A white vapor was observed over the pot
After 15 minutes, the pot was one quarter of the way full

a. Some of the water disappeared
b. The water changes from a gas to a liquid
c. Some of the water changed from a liquid to a gas, which was the vapor observed
d. The pot was never half full, someone measured incorrectly

28. A ball is resting on the front end of a boat. The boat is moving straight forwards toward a dock. When the front of the boat hits the dock, how will the ball's motion change?

a. The ball will remain at rest
b. The ball will move backwards
c. The ball will move forwards
d. The ball will move sideways

29. Which statement describes how fossil fuels are made?

a. Oil from Earth's core slowly rises to the Earth's surface and forms ponds
b. Lava within some volcanoes becomes so hot that is transforms into fossil fuels
c. Dead and decomposing organisms, exposed to extreme heat and pressure, transform into fossil fuels
d. When mountains form on top of coal, the pressure liquefies the coal creating fossil fuels

30. The Sahara desert receives about 3.6 inches of precipitation per year. Antarctica is also a desert receiving between 3 to 8 inches of precipitation per year. What makes an area a desert?

a. The temperature
b. The amount of animal life
c. The number of people living in the region
d. The yearly precipitation

31. After examining the two diagrams, the student made some observations. Which observation is NOT correct?

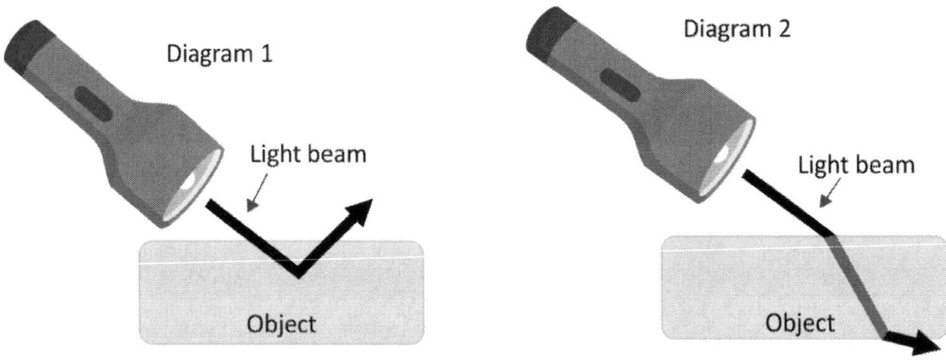

a. Diagram 1 is an example of how light is reflected.
b. The object in diagram 1 could be a wooden block.
c. Diagram 2 is an example of how light is refracted.
d. The object in diagram 2 could be water.

32. A prairie ecosystem is described below. Which organisms from the ecosystem are carnivores?

Grass is eaten by grasshoppers, rabbits, and crickets
Corn is eaten by squirrels and beetles
Grasshoppers, beetles, and crickets are all eaten by birds
Squirrels, birds, and rabbits are all eaten by both foxes and coyotes

a. Corn and grass
b. Grasshoppers and crickets
c. Beetles and birds
d. Foxes and coyotes

33. In a food chain, where does energy go after the secondary consumer dies?

a. Back to the Sun
b. To the producers
c. Into air, becomes wind
d. To decomposers

34. Every year thousands of square miles of trees are cut down to make paper products. Which of the following is not a result of cutting down trees?

a. Animals experience a loss of habitat
b. There is a decrease in the amount of oxygen available
c. There is a decrease in the amount of food available to herbivores and omnivores
d. There is a decrease in the amount of CO_2 in the atmosphere

35. According to the data below, what can be determined about Saturn?

Planet	Length of Year	Length of Day
Venus	224.7 days	116.75 days
Earth	365 days	24 hours
Saturn	10,759 days	10 hours 32 minutes
Neptune	164.79 years	16.11 hours

a. Saturn takes the least amount of time to rotate on its axis.
b. Saturn is the furthest planet from the Sun.
c. The further a planet is from the Sun, the shorter its day will be.
d. The larger a planet is, the longer its day will be.

36. A dog is an omnivore. What would you expect a dog's diet to consist of?

a. Meat
b. Vegetables
c. Both meat and vegetables
d. Neither meat nor vegetables

37. The following represents a simple food chain. What trophic level contains the greatest amount of energy?

tree → caterpillar → frog → snake → hawk → worm

a. tree
b. caterpillar
c. hawk
d. worm

38. How are organisms, such as snakes, cacti, and coyotes, able to survive in harsh desert conditions?

a. Over thousands of years these organisms have developed adaptations to survive in arid climates
b. These organisms migrate out of the desert during the summer months, only living in the desert for a portion of the year
c. Snakes, cacti, and coyotes work together to find sources of food and water
d. Snakes, cacti, and coyotes are all aquatic species that live in ponds and rivers during the hot day

39. A student is building a model of the Solar System using produce from the local supermarket. What order should she put the objects in to correctly represent the Solar System?

Object in Solar System	Model	Object in Solar System	Model
Mars	Raspberry	Uranus	tomato
Saturn	Grapefruit	Mercury	raisin
The Sun	Watermelon	Venus	grape
Jupiter	Lettuce	Asteroid Belt	sunflower seeds
Earth	Cherry	Neptune	orange

a. Watermelon-raisin-grape-cherry-raspberry-tomato-orange-lettuce-grapefruit-seeds
b. Watermelon-raisin-grape-cherry-raspberry-seeds-lettuce-grapefruit-tomato-orange
c. Watermelon-lettuce-grapefruit-tomato-orange-cherry-grape-raspberry-raisin-seeds
d. Watermelon-grape-raisin-cherry-seeds-raspberry-lettuce-grapefruti-orange-tomato

40. Choose from the given types of energy, and next to each item below, write the type of energy needed for the item to perform its main function. Select only one choice for each object. Each type of energy may be used more than once.

Mechanical Energy	Electrical Energy	Light Energy

Bicycle:____________

Sun:____________

Toaster:____________

Hair dryer:____________

Lit candle:____________

41. A student participated in a science fair. He was asked to complete an experiment that would test the effect of force on an object. He completed an experiment in which he rolled a baseball on a turf field, a grassy field, a rocky path, and an ice rink and timed how long it took the baseball to go the same distance. What question was he investigating?

a. What types of objects move with less force?
b. What is the effect of friction on moving objects?
c. What types of objects roll more easily?
d. What is the effect of mass on the force needed to move an object?

42. Which of the following is considered a non-renewable resource?

a. Glass
b. Wood
c. Cattle
d. Soil

43. How is force being applied to the box below?

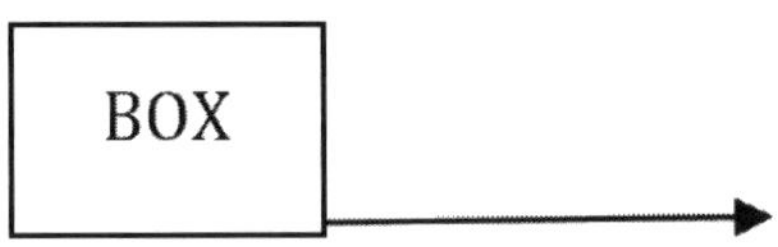

a. The box is being pulled forward
b. The box is being pushed forward
c. Gravity is forcing the box to move forward
d. Friction is forcing the box to stop

44. A student completed a science lab in which he had to shine a flashlight on three different objects. He then completed a chart based on his observations. Which chart below has been completed correctly?

a.

Object	Light Beam
Glass of water	Reflected
Flat mirror	Refracted
Ice crystals	Scattered

b.

Object	Light Beam
Glass of water	Refracted
Flat mirror	Reflected
Ice crystals	Scattered

c.

Object	Light Beam
Glass of water	Reflected
Flat mirror	Scattered
Ice crystals	Refracted

d.

Object	Light Beam
Glass of water	Scattered
Flat mirror	Reflected
Ice crystals	Refracted

45. Which of the light bulbs below will light up in the circuit when the switch is left open, as shown below?

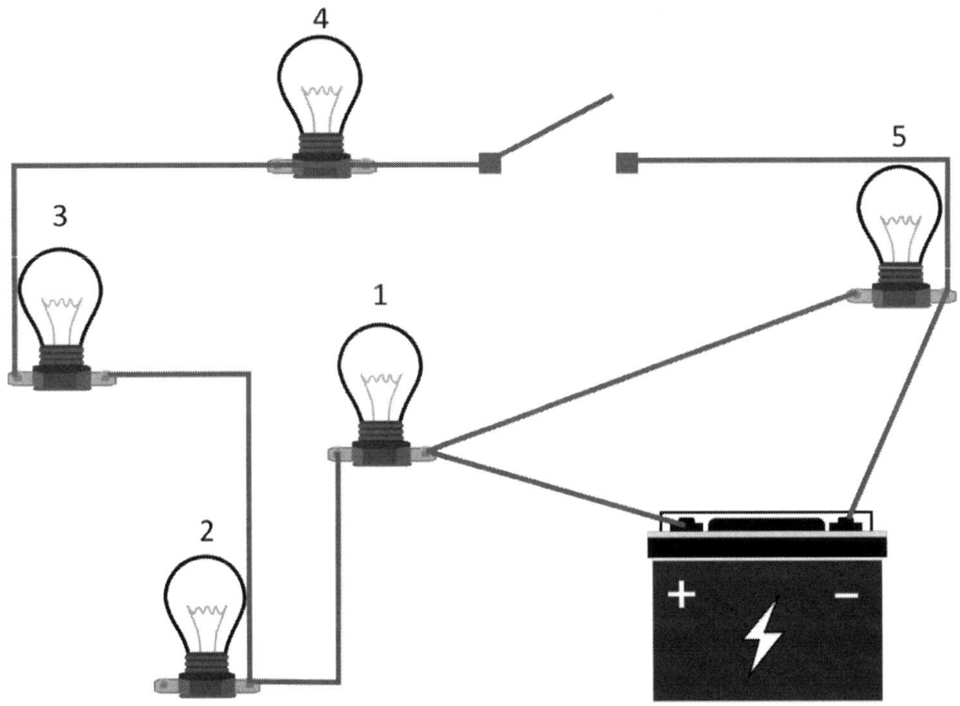

a. Light bulb 1 only
b. Light bulbs 1 and 5
c. Light bulbs 1, 3, and 5
d. Light bulb 5 only

46. A student put some water in a small bird bath in the garden at her school. A week later she had to add more water because it had evaporated. What type of energy caused the water in the bird bath to evaporate?

a. Thermal energy from the Sun
b. Sound energy from the chirping birds
c. Light energy from the nearby lamp post
d. Mechanical energy from the pinwheels in the garden

47. Fill in the bubble for each buzzer in the circuit that will sound if the switch is left open. Choose only THREE buzzers.

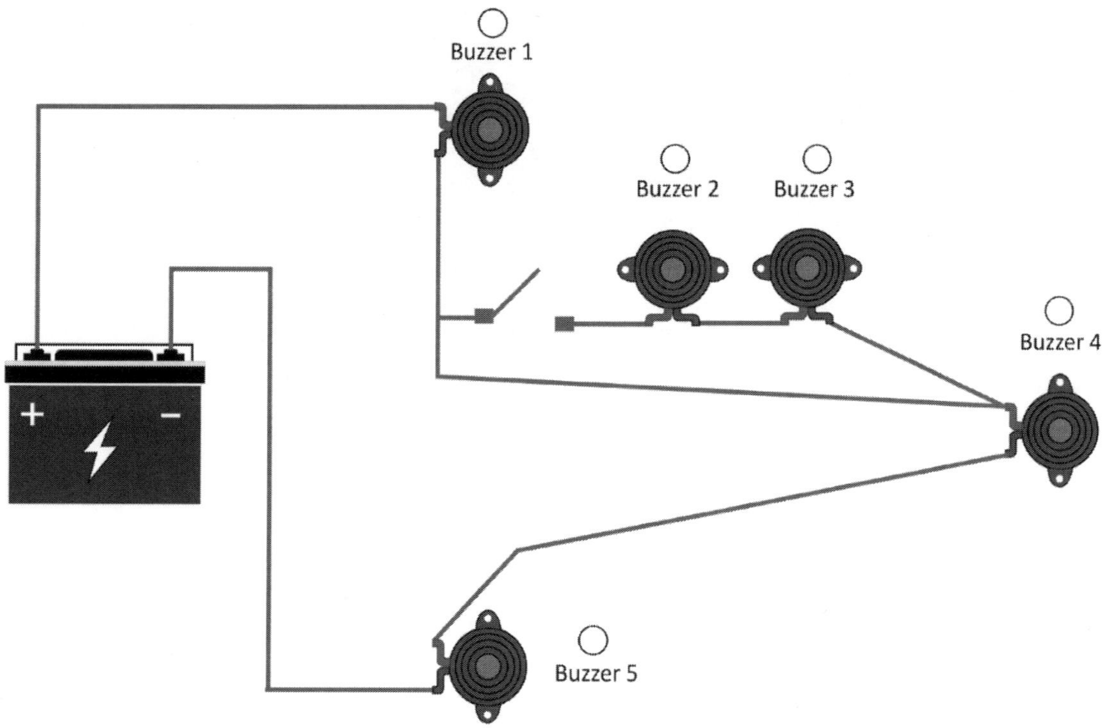

48. A student is writing a paragraph explaining the steps that occur when sedimentary rock is forming. The student has written the outline below. Using the options below, determine which responses belong in steps 4 and 5. Only two choices will be used.

1.	**Weathering:** This occurs when rocks are broken down by things such as wind and water.
2.	**Erosion:** This occurs when the pieces of rock that have been weathered are moved away.
3.	**Deposition:** This occurs when the sediments are laid down.
4.	
5.	

Match the correct options to the empty spaces in the table above.

Compaction: As more sediment is deposited, the pieces of sediment are pushed closer together.	**Cementation:** The sediment that has been deposited is bound together to form a new layer of rock.
Cementation: The sediment breaks into smaller pieces until there is no space left between any of the pieces.	**Compaction:** Sediment continues to settle until the pieces of sediment make an even layer.

49. The student completed a science lab and drew this diagram to explain what happened.

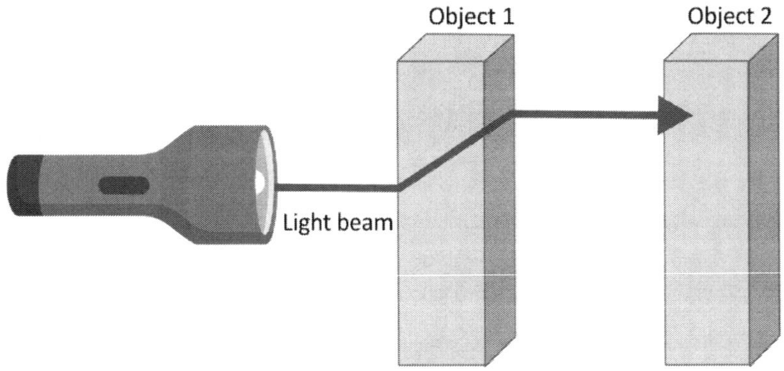

Below the diagram, he wrote five observations. Based on the diagram, which observation is false?

1. Light travels in a straight line until it strikes an object.
2. The light beam was refracted as it traveled through object 1.
3. The light beam was absorbed by object 2.
4. Object 1 was a container of water.
5. Object 2 was a mirror.

a. Only observation 1 is false because light can change directions even without striking another object.
b. Only observation 2 is false because the light was reflected when it encountered object 1.
c. Only observation 4 is false because both objects must be solids.
d. Only observation 5 is false because the light was absorbed by object 2, and a mirror would reflect the light.

50. The science class worked together to make a list of the types of energy used by various objects. Below is an example of one list.

Object: Microwave

Type of Energy Used	Example of Energy Use
Sound	A beeper sounds when the object has finished its task.
__________	There is a light bulb inside the object to illuminate what is inside.
__________	The object warms the food.
__________	The door of the object opens when a person pulls on the handle.
Electrical	Electricity is used to power the object.

Match the correct type of energy to each location in the table. Not all answer choices will be used.

Thermal	Electrical	Sound	Mechanical	Light

Answer Key and Explanations for Test #1

1. B: Thermometers are used to measure temperature. Graduated cylinders are used to measure volume while a ruler is used to measure distance. Balances are used to measure mass.

2. A: A butterfly goes through the same stages as a house fly. It starts with an egg and progresses to larva or caterpillar, then the pupa or chrysalis, and then an adult. After being born from an egg, a frog goes from a tadpole to froglet to an adult frog (B). A bird hatches from an egg and grows from a baby bird to an adult bird (C). A bat is born live, not from an egg, so its life cycle is very different from the house fly (D).

3. B: The east coast, states such as Maine, New York, and New Hampshire are three hours ahead of states on the west coast such as California, Washington, and Oregon. Earth rotates west to east, and that is counterclockwise when viewed from above the North Pole.

4. Answers may vary: The scientist can conclude that the animal ate meat. Since sharp teeth would be more helpful for chewing meat, the scientist is led to believe that this animal ate meat because the teeth are sharp and pointy. These teeth also appear to come from a large animal, but it is not clear if these teeth came from one animal or several. The only reasonable claim for the scientist to make based off of the information given is about what type of meat the animal ate.

5. A: Heat on Earth is generated by the Sun. The more direct sunlight an area on Earth receives from the Sun, the warmer it will be. When the Northern hemisphere is tilted away from the Sun, all of the countries in the Northern hemisphere experience winter. At that same time, the Southern hemisphere experiences summer. The same is true when the Southern hemisphere experiences winter; the Northern hemisphere experiences summer.

6. A: The students are trying to increase the speed of the car without physically pushing it with more force, so they are using an incline plane. Through this experiment, the students should discover that raising the level of the inclined plane increases the speed of the car. The car's weight (B) and size (C) were never changed, and the exact amount of force was never measured (D). Because of this, these questions could not be answered by the experiment.

7. D: According to the description of the poison dart frog, the warm and rainy temperature of ecosystem T would be the best for it. Both the very cold climate found in ecosystem Q (A) and the very hot climate found in ecosystem R (B) would likely be too extreme for the frog. The plains, high summer temperatures, and lack of flowing water in ecosystem S (C) would not benefit the frog. The temperatures, streams, rivers, trees, and lush land found in ecosystem T make it a good habitat for the dart frog.

8. D: Igneous rocks are formed when magma in the Earth erupts through cracks in the crust. There, the lava cools, creating a hard structure with many air pockets or holes.

9. B: Algae are a type of plant that grows in water. Because algae are plants, they are able to capture light from the Sun and transform it into energy during the process of photosynthesis. Algae do not eat other organisms (C), so they do not need to hunt for fish (A). Fossil fuels are buried deep within the Earth's surface. Algae could not reach them, and algae do not need to consume them (D).

10. D: The circuit pictured on option D represents a closed circuit when the switch is closed because the closed switch and conducting wires connect one side of the battery to one side of the propeller motor and the opposite side of the battery to the opposite side of the propeller motor. This allows the electrical current to travel successfully. Option A would not cause the propeller to spin when the switch is closed because only one side of the propeller motor is connected to the power source. Option B would not cause the propeller to spin when the switch is closed because the wires are only connected to one side of the battery. Option C would not cause the propeller to spin when the switch is closed because the wire only connects to one side of the switch, which makes it an open circuit.

11. C: The prefix *carni-* means flesh, so carnivores are flesh eaters. Their diet consists of meats. The only answer choice that is a meat is an animal.

12. A: Plants are considered producers because they are able to absorb light energy from the Sun and convert it into food through the process of photosynthesis. In other words, plants are producers because they produce their own food from sunlight.

13. A: The volume of a graduated cylinder should be read at the bottom of the meniscus.

14. D: The planets ordered from closest to farthest from the Sun are Mercury, Venus, Earth, Mars, Jupiter, Saturn, Uranus, and Neptune. Option D lists the planets in the correct order. Option A is incorrect because it lists Jupiter before Venus and Earth. Option B is incorrect because it lists Saturn before Jupiter. Option C lists the planets from farthest to closest to the Sun, so it is incorrect.

15. D: In the water cycle, evaporation is the process of water being heated by the Sun and turning into vapor, which is a gas. Condensation is the process of water vapor in the air being cooled and turning into tiny droplets of water. If the vapor were heated, it would continue to be vapor. Precipitation is the process of water falling to the surface of the Earth either as a liquid or solid. Collection is the process of fallen precipitation flowing downhill and collecting in rivers, lakes, and oceans. Water does not naturally flow uphill.

16. D: The graph represents a set of data. The line graph increases in distance from 1 second to 4 seconds and then begins decreasing. The only data set that shows distance increasing and then decreasing is set D.

17. A: The weather forecast shows that the temperature will drop well below 32°F overnight and stay in the 20s. The freezing point of water is 32°F, which means that the water in the birdbath will freeze overnight after the temperature drops. Choice B is incorrect because the water in the birdbath would need to get very hot in order to evaporate, which will not happen, according to the forecast. Choice C is incorrect because the water in the birdbath would need to be frozen and then heated in order to melt. Choice D is incorrect because as the water in the birdbath freezes, it will contract and it will appear that there is less than there was. This will not cause the birdbath to overflow.

18. A: The microscope was discovered when a series of lenses were placed in a specific order magnifying objects that were too small to see otherwise. The development of the microscope also led to discoveries such as the cell, bacteria, and fungi.

19. B: The purpose of the rubber coating on the outside of electrical cords is to contain the electric current in the metal wire. When part of the rubber coating is missing, the electric current can leave the wire and cause an electric shock to the person plugging in the wire.

20. A: When the Earth completes one rotation on its axis, it is equal to one complete day, or approximately 24 hours. If the rotation of the Earth happened more quickly, a day would be shorter than 24 hours. The time it takes the Earth to rotate on its axis would not affect the seasons (B). Years are based on the amount of time it takes the Earth to orbit the Sun. While this change would add more days to a year, it would not make the year shorter (C). The Earth rotating faster wouldn't keep night from happening (D). It would just make both day and night shorter.

21. D: The light bulb will not light up. For the circuit to work correctly, both ends of the battery (the energy source) need to be connected by different connecting wires. In this picture, the bottom of the battery is not connected to a wire, so the circuit will not work and, the bulb will not light up (D). The circuit is not closed (C), so it will not light the bulb, even though the bulb is connected by the wire (A). The fact that the wire is diagonal does not affect whether or not the circuit will light the bulb (B).

22. D: The molecules that make up gases are far apart and move about very quickly. Because they have a weak attraction to each other, they expand to take up as much space as possible.

23. A: Light always travels in a straight line, and it cannot move around an object that gets in its way. The light from the projector traveled in a straight line to the screen. When the student stood directly in front of the projector, the light was blocked because it could not move around the student. Light is scattered when it is fully absorbed by an object and that object reflects the light back out. This often happens when light encounters thin clouds or ice crystals, but it does not always happen when light hits a solid object (B). The light would not have been reflected off the student's shirt, since it does not have a reflective surface (C). The light would not move upwards just because another object got in the way (D). The object would need to have a reflective surface in order to reflect the light upward.

24. D: Mechanical energy is used when something is moved. The carpenter must move the hammer, drill, wrench, and saw in order to use them, so the carpenter needs to use mechanical energy. Although two of the tools are electrical, the hammer and the wrench are not, so only two of the tools require electrical energy (A). The tools do not require heat (B) or light (C) in order to be used.

25. B: Light travels faster in air than it can in water. Water molecules are closer together than air particles are, which causes the light to slow down and bend as it enters the water. The bending of light is called refraction and creates the illusion of the wallet being next to where it actually is.

26. A: In this experiment, erosion occurred when the water removed small pieces of dirt from part of the mountain and moved them to a new place. Melting (B) involves heat, which is not present in this experiment. Decaying (C) involves rotting, which is not happening in this experiment. Cementation (D) occurs when dissolved minerals bind pieces of sediment together and eventually form new rock. That is not happening in this experiment, either.

27. C: Water boils at 100°C. As the pot of water heats up over the campfire, it begins to boil causing the liquid water to turn into water vapor, a gas. The vapor the campers observed over the boiling pot of water was water vapor. As the water continues to boil, more of the liquid turns into gas resulting in a decrease in the volume of water in the pot.

28. C: The ball will move forwards. The ball is moving forward with the boat. When the front of the boat hits the dock, the ball's motion does not change. It continues to move forward because the force acting to stop the boat is not acting upon the ball. The forward motion of the boat is halted by the dock. The forward motion of the ball is not stopped. Since the ball is round there is little friction to provide an equal and opposite reaction to the forward motion.

29. C: Organisms that lived millions of years ago have been covered up by many layers of earth. The temperature and pressure increases transforming the dead organisms into fossil fuels that can be obtained and used for energy.

30. D: A desert is classified by the average amount of precipitation the region receives on a yearly basis. The temperature is irrelevant, as there are both hot and cold deserts. Due to the dry conditions, only organisms designed to withstand the arid conditions can survive in these areas.

31. B: Choice B is the incorrect observation. Diagram 1 illustrates how light is reflected, and the object is probably a mirror or other reflective object (A). Object 1 could not be a wooden block because wood does not reflect light (B). Diagram 2 is an example of how light is refracted (C) when it moves from air to a liquid, like water (D).

32. D: Carnivores eat meat only. In the ecosystem described, both coyotes and foxes are carnivores eating squirrels, birds, and rabbits.

33. D: After a secondary consumer dies, such as a wolf, its body is partially consumed by decomposers, such as bacteria and fungi. Bacteria and fungi live in soil and digest body tissues of dead organisms converting them into basic nutrients that plants need to grow. Therefore, after secondary consumers die, their energy is consumed by decomposers, who make nutrients available in the soil for producers to use. *Do not confuse nutrients in the soil with energy that producers get from the Sun to make their own food.*

34. D: Trees absorb and use carbon dioxide to produce their own food. Humans breathe in oxygen and breathe out carbon dioxide. To a human, too much carbon dioxide is toxic and can cause serious problems. Plants use carbon dioxide and release oxygen as waste. This allows what is toxic to humans, carbon dioxide, to be used and transformed into breathable oxygen. A decrease in the number of trees decreases the amount of plants taking in carbon dioxide, causing the carbon dioxide levels in the atmosphere to rise.

35. A: When a planet rotates on its axis it creates day and night. The side of the planet that faces away from the Sun is in night. When that side rotates around to face the Sun, it is in day. The data indicates that Saturn takes the least amount of time to rotate on its axis, which indicates that is has the shortest day at only 10 hours and 32 minutes.

36. C: The prefix *omni*-means all, so an omnivore eats all types of food. Omnivores, such as dogs and many humans, eat a variety of plants and meats to achieve a balanced diet.

37. A: In the food chain of tree → caterpillar → frog → snake → hawk → worm, the tree is at the trophic level with the greatest amount of energy. Trophic level refers to the position of an organism in a food chain. Energy is lost according to the laws of thermodynamics as one moves up the food chain because it is converted to heat when consumers consume. Primary producers, such as autotrophs, are organisms who are at the base and capture solar energy. Primary consumers are herbivores that feed on the producers. Secondary consumers consume primary consumers and so on. Decomposers get their energy from the consumption of dead plants and animals.

38. A: Many organisms, especially organisms that live in harsh conditions such as deserts or frozen icy areas, have developed specific adaptations that allow them to survive. For example, cacti are able to expand to store large amounts of water, coyotes absorb some water from their food, and snakes can escape the heat by hiding within rocks.

39. B: This question is really asking you to determine the correct order of the major bodies in the Solar System. The correct order is: Sun (watermelon), Mercury (raisin), Venus (grape), Earth (cherry), Mars (raspberry), asteroid belt (seeds), Jupiter (lettuce), Saturn (grapefruit), Uranus (tomato), Neptune (orange).

40.

Mechanical Energy	Electrical Energy	Light Energy
Bicycle:	Mechanical Energy	
Sun:	Light Energy	
Toaster:	Electrical Energy	
Hair dryer:	Electrical Energy	
Lit candle:	Light Energy	

Because a bicycle has moving parts that must be moved for it to perform its main function, which is transportation. Mechanical energy is required for these parts to make the person riding the bicycle move. Electrical energy is energy produced by moving electrical charges. A toaster and hair dryer both require electrical energy to perform their main functions. Light energy is energy that comes from light waves, some of which we can see. The Sun and a lit candle both use light energy to perform their main functions.

41. B: Friction is the force that resists an object from sliding or rolling. The student was testing which surfaces had more resistance than others when rolling a ball. He was investigating the force of friction. The type of object was a constant throughout each experiment (B), and the mass of the object moving did not change (D). Since these factors did not change during the experiment, options A, C, and D are not questions the experiment would answer.

42. A: Glass is considered a non-renewable resource. Glass is manufactured and can be recycled, but is considered a non-renewable resource. Answer B, Wood, is considered a renewable resource because with proper management, an equilibrium can be reached between harvesting trees and planting new ones. Cattle, Answer C, are managed in herds and a balance can be achieved between those consumed and those born. Answer D, Soil, is the result of long-term erosion and includes organic matter and minerals needed by plants. Soil found naturally in the environment is renewed. Crops can be rotated to help maintain a healthy soil composition for farming.

43. A: The picture shows an arrow in front of the box pulling the box forward. If the box were being pushed, the arrow would be behind the box pushing it forward. Gravity pulls downward, not forward.

44. B: When light moves from air to water, it is refracted and appears to bend. When light hits a mirrored object, it is reflected off the surface of the mirror. When light encounters ice crystals, the light is scattered. The light is first absorbed by the ice crystal, but then the light is sent back out through the crystal with weaker strength. Only choice B correctly lists each effect.

45. D: Light bulb 5 is the only light bulb that is part of a completely closed circuit when the switch is open. Light bulbs 1, 3, and 4 are part of an open circuit because when the switch is open, they are only connected to one side of the battery. If the switch were closed, those light bulbs would light up as well. Only one side of light bulb 2 is connected to the circuit, so it is not part of a closed circuit even when the switch is closed. Therefore, it will not light up if the switch is open or closed.

46. A: Evaporation is caused by heat. The Sun gives off a high amount of thermal energy, so it can be concluded that thermal energy from the Sun caused water to evaporate from the bird bath. Neither sound energy from birds (B) nor mechanical energy from pinwheels (D) would produce heat, so neither would cause evaporation. Although a light bulb may produce a small amount of heat, it would not produce enough heat to cause this amount of water to evaporate (C).

47.

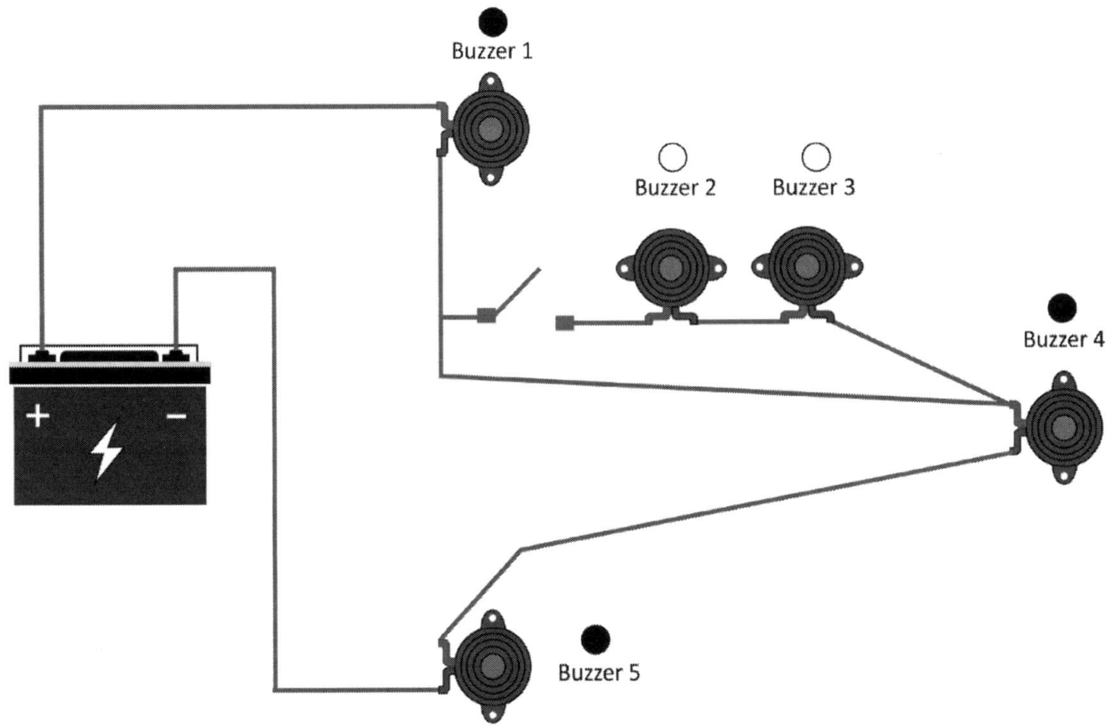

Buzzers 1, 4, and 5 are part of a closed circuit because the wire connected to the battery runs in a complete path that is connected to each of those buzzers. This creates a closed circuit. Since the switch is in the open position, buzzers 2 and 3 are not part of the closed circuit, so the electric current does not travel to these buzzers. If the switch were closed, all of the buzzers would sound.

48.

1.	**Weathering:** This occurs when rocks are broken down by things such as wind and water.
2.	**Erosion:** This occurs when the pieces of rock that have been weathered are moved away.
3.	**Deposition:** This occurs when the sediments are laid down.
4.	**Compaction:** As more sediment is deposited, the pieces of sediment are pushed closer together.
5.	**Cementation:** The sediment that has been deposited is bound together to form a new layer of rock.

The steps to form sedimentary rock follow the following pattern: weathering, erosion, deposition, compaction, and cementation. Compact means tightly packed together, so the pieces of sediment are pushed closer together during compaction. The process of cementation is when a new layer of rock is formed. When more sediment is deposited on top of other sediment, the sediment that was there before is pushed down and the space between the pieces shrinks. Sediment does not always settle in an even layer. Sediment also does not always continue to break down.

49. D: Observation 5 is the only false statement. According to the diagram, object 2 absorbed the light beam. If object 2 was a mirror, the light would have been reflected (D). Observation 1 is correct because light always travels in a straight line until it strikes an object (A). Observation 2 is correct because the diagram shows the light beam being refracted as it passes through object 1 (B). Observation 3 is correct because according to the diagram, object 2 absorbed the light beam. Observation 4 is correct because a container of water would refract the light as shown in the diagram (C).

50.

Type of Energy Used	Example of Energy Use
Sound	A beeper sounds when the object has finished its task.
Light	There is a light bulb inside the object to illuminate what is inside.
Thermal	The object warms the food.
Mechanical	The door of the object opens when a person pulls on the handle.
Electrical	Electricity is used to power the object.

The light emitted by the bulb in the microwave is an example of light energy. Thermal energy is used to warm the food in the microwave. Opening the microwave door requires mechanical energy because the door must be moved.

Practice Test #2

Refer to the following for questions 1-2:

A prairie ecosystem is described below. Farmers treat fields surrounding the prairie with pesticides that kill part of the beetle population.

1. What effect would the pesticides have on the rest of the ecosystem?

Grass is eaten by grasshoppers, rabbits, and crickets
Corn is eaten by squirrels and beetles
Grasshoppers, beetles, and crickets are all eaten by birds
Squirrels, birds, and rabbits are all eaten by both foxes and coyotes

a. There would be a decrease in the cricket population.
b. There would be an increase in the population of birds.
c. The population of rabbits would decrease.
d. The population of foxes would increase.

2. Which organisms from the ecosystem described are herbivores?

a. Corn and grass
b. Grasshoppers and crickets
c. Beetles and birds
d. Foxes and coyotes

3. Which statement correctly describes the relationship between plants and animals?

a. Animals breathe in oxygen and plants release oxygen.
b. Animals breathe in carbon dioxide while plants release carbon dioxide.
c. Animals breathe in carbon dioxide while plants breathe out carbon dioxide.
d. Animals get oxygen from plants by eating the plants.

4. Which picture best represents Earth's position on its axis?

a. 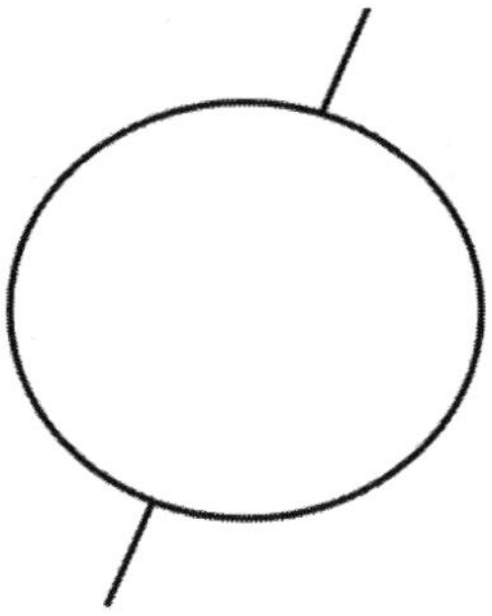

b. 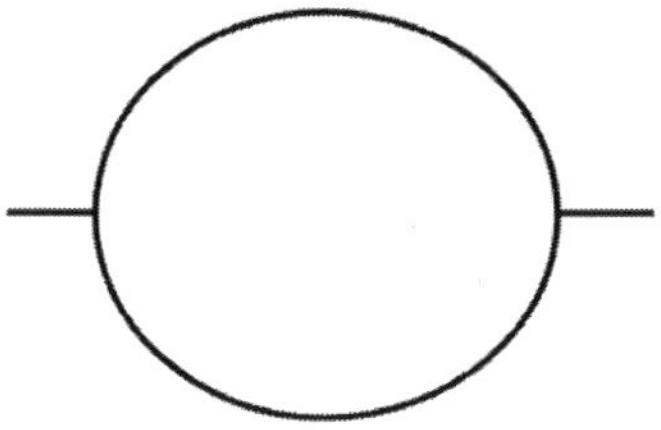

c. 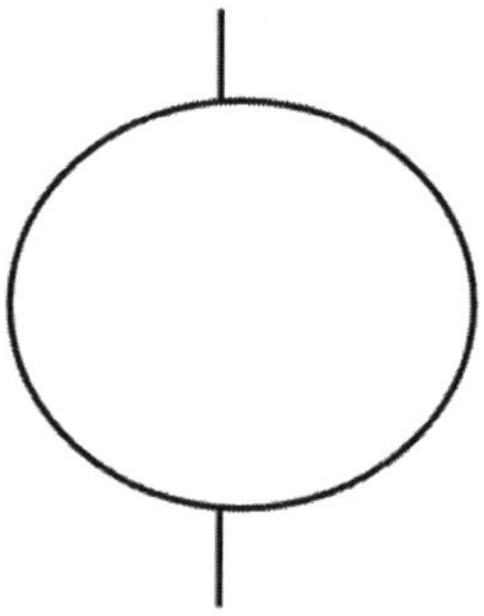

d. 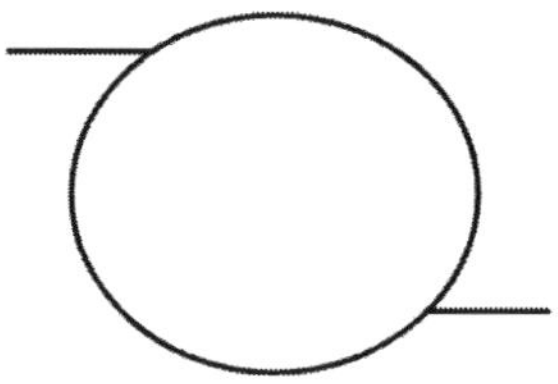

5. What should be included in a hypothesis?

a. A hypothesis should tell what you think is going to happen and why.
b. A hypothesis should state what the results of similar experiments were.
c. A hypothesis should summarize the results of an experiment.
d. A hypothesis should compare the control to the variable.

6. What causes high and low tide?

a. The Earth's rotation
b. The Moon's gravity
c. The wind currents
d. The change in water temperature

7. A student wrote the following paragraph in science class. Which of the student's statements about the Earth and the Sun is NOT correct?

The Sun looks like it moves across the sky during the day, but it actually doesn't move at all. It just looks this way because the Earth is spinning on its axis. It takes about 48 hours for the Earth to rotate one full time around its axis. The Earth is also moving in an orbit around the Sun.

a. The Sun looks like it moves across the sky during the day, but it doesn't actually move.
b. The Earth spins on its axis.
c. It takes about 48 hours for the Earth to rotate one full time on its axis.
d. The Earth moves around the Sun.

8. What does a food chain show?

a. The flow of energy between organisms
b. Every type of species living in a habitat
c. The population of each species in a habitat
d. The number of offspring a species will produce in a year

9. Lions live in savannah and grassland regions where there are tall dry grasses. Lions must sneak up and stalk their prey. What is the most likely reason for a lion's sand colored coat?

a. It allows lions to hide within the grass
b. The light color of their coat reflects heat from the Sun
c. A lion's coloring helps it attract mates
d. It allows them to run faster than their prey

10. The picture below is of a cast-iron skillet. The skillet is sold with a rubber handle cover. Why would the manufacturer sell it with a rubber sleeve to cover the handle?

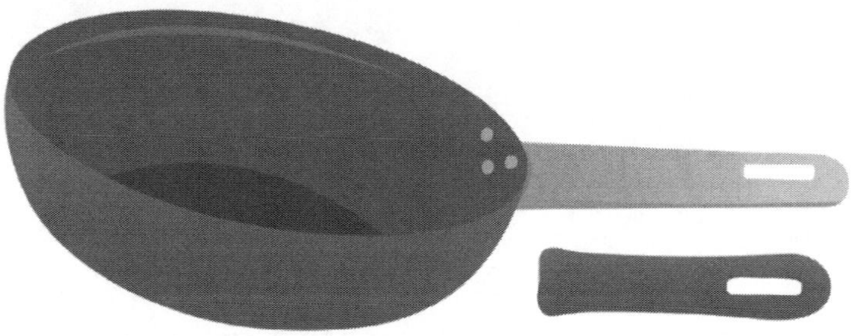

a. Rubber is denser than water. When it is time to wash the pan, the owner should add the rubber sleeve to the handle so that the pan will sink in the water. This will make it much easier to wash.
b. Since iron is a good conductor of thermal energy, the iron handle on the pan will get hot quickly and could burn the user. Since rubber is an insulator, when the rubber sleeve is placed over the iron handle, the handle will not get as hot and will not burn the user.
c. Both iron and rubber are good conductors of thermal energy. When the rubber sleeve is placed over the iron handle it will help the pan heat up more quickly, and the food will take less time to cook.
d. Rubber is a good conductor of thermal energy. When the rubber sleeve is added to the handle, it will keep the food in the skillet warmer for a longer period of time.

11. Wind farms are one source of alternative energy, energy that can be used in place of fossil fuels. The amount of energy a wind farm produces is determined by

a. Sunlight
b. Season
c. Wind
d. Rainfall

12. A carnivore would eat all of the following except

a. Rabbits
b. Acorns
c. Fish
d. Deer

13. What instrument should be used to measure how fast a boy can run 100 meters?

a. Beaker
b. Meterstick
c. Stopwatch
d. Thermometer

14. A student woke up early to iron his shirt for school. After getting dressed, he put some waffles in the toaster and then went to the bathroom to use the hair dryer to dry his hair.

Which best describes the two main types of energy that are being used by all of the objects to help accomplish their main functions?

a. Sound energy and light energy
b. Mechanical energy and sound energy
c. Light energy and electrical energy
d. Electrical energy and thermal energy

15. What would be the best material to make a cooking pot out of?

a. Rubber
b. Copper
c. Plastic
d. Cement

16. Flowers, trees, and shrubs are all examples of what?

a. Carnivores
b. Herbivores
c. Producers
d. Predators

17. Why can a person see their reflection in a mirror?

a. Light is scattered when it hits a mirror producing a backward image
b. Light is reflected off of the mirror back towards the person standing in front of it
c. Light bends as it passes through the mirror creating an optical illusion
d. Light is refracted, which produces a mirror image behind the glass

18. A tropical flower is planted in a local garden. The garden is watered 2 to 3 times per week during all seasons. After only a few months the tropical plant begins to droop and turn brown. How did environmental change most likely affect the tropical plant?

a. Being watered 2 to 3 times per week was not enough to sustain a tropical plant
b. The new environment was too hot and burned the plant
c. Insects attacked the plant causing it to wilt
d. The new soil lacked the nutrients necessary to maintain the plant's health

19. Place your answer on the provided griddable answer sheet.

How many planets are in Earth's solar system?

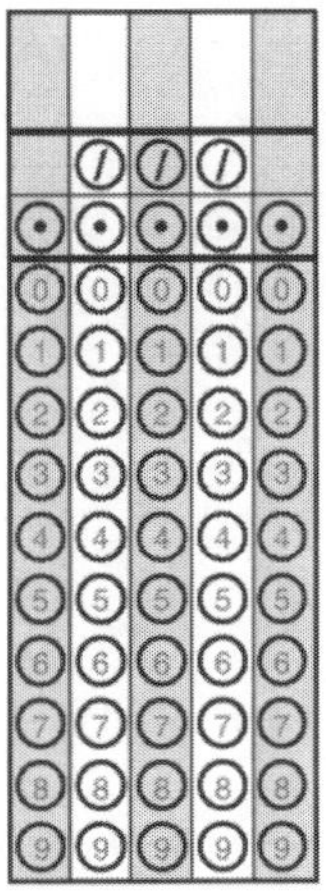

20. A student saw a model of the bones of this dinosaur at a natural history museum.

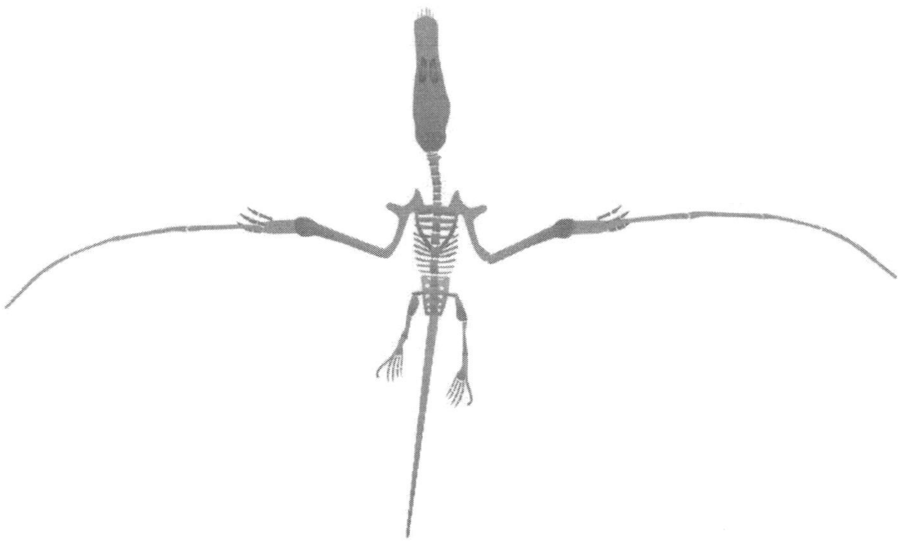

What question could the student most likely answer by looking at the skeleton of this dinosaur?

a. What type of food did the dinosaur eat?
b. During what time period did the dinosaur live?
c. How did the dinosaur get from one place to another?
d. Did the dinosaur travel in a pack or alone?

21. Algae are part of many food chains. How do algae produce energy?

algae fish seal shark

a. They hunt fish for food
b. Algae undergo photosynthesis
c. They decompose dead or dying organisms
d. They consume fossil fuel from beneath Earth's surface

22. Four windows are washed with different cleaners. The same paper towel is used on each. The squares below show each window before being cleaned. What is the variable in the experiment?

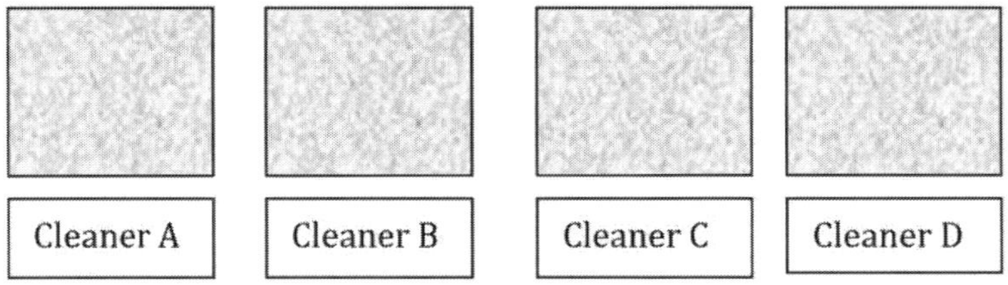

a. The amount of sunlight coming through the windows
b. The type of window cleaner used
c. The amount of dirt on each window
d. The type of paper towel used

23. Why is it more difficult to push a shopping cart full of groceries than an empty shopping cart?

a. The full cart has less mass than the empty cart
b. The full cart has a greater mass than the empty cart
c. The full cart has less friction than the empty cart
d. The empty cart is not pulled down by gravity

24. Use the information in the table to determine what would happen to the mass and weight of a human if he or she were on Neptune.

Mass	Mass is the amount of matter there is. Mass only changes when matter is added or removed.
Weight	Weight is how much gravity pulls downward on an object. The more gravity there is, the more an object weighs.
Gravity on Earth: 9.8 m/s^2	Gravity on Neptune: 11.15 m/s^2

a. A person's mass and weight would increase on Neptune
b. A person's mass and weight would decrease on Neptune
c. A person's mass would increase, but weight would be unchanged
d. A person's mass would be unchanged, but weight would increase

25. This picture shows a college baseball player hitting a baseball with a baseball bat.

Which choice below correctly shows how the position of the ball will change and gives the correct reason for this change?

a.

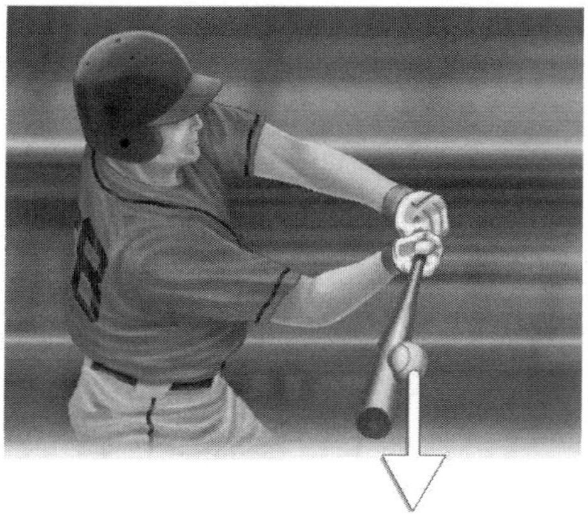

The force of gravity forces the ball to fall.

c.

The force of the bat pushes the ball away from the batter.

b.

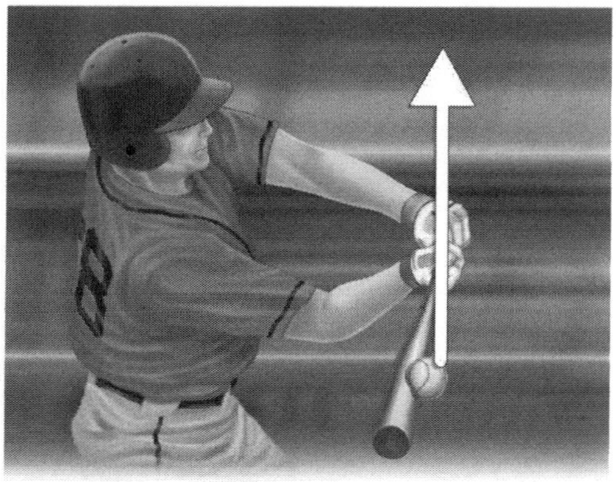

The force of the bat will pull the ball straight up.

d.

The force of the bat pulls the ball away from the batter.

26. How are the size of a dog and the pitch of its bark related?

Breed	Height from ground to back	Pitch of Bark
Chihuahua	20 cm	17 Hz
Cocker Spaniel	45 cm	14 Hz
Labrador	63 cm	12 Hz
Dalmatian	78 cm	11 Hz
Great Dane	114 cm	8 Hz

a. The smaller breeds have a lower pitch bark than the larger breeds
b. The size of the dog has no relationship to the pitch of the bark
c. The larger breeds have a lower pitch bark than the smaller breeds
d. The smallest breed and largest breed have the highest pitch barks

27. How are we able to see the Moon on a dark night?

a. The Moon reflects light from the Sun, which illuminates the Moon
b. The Moon generates energy that makes it glow
c. The Moon reflects heat from the Sun that causes it to glow
d. The Moon's surface is covered in active volcanoes that glow when they erupt

28. Below is a diagram of the life cycle of a frog and the life cycle of a sunflower.

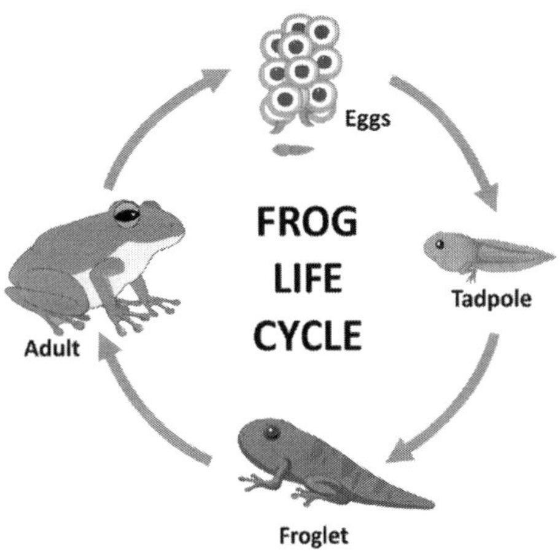

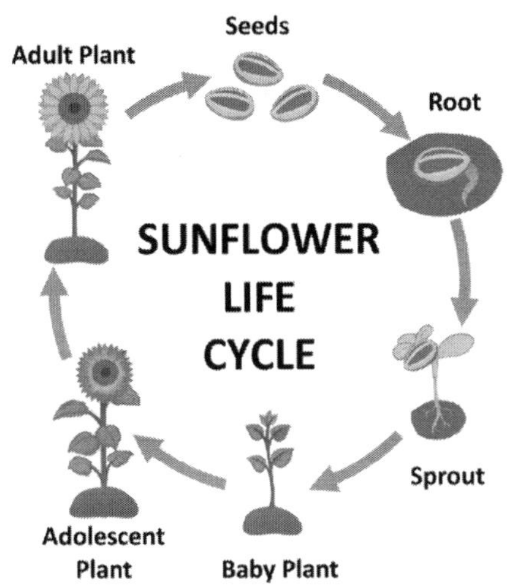

What are TWO similarities between the life cycle of a frog and the life cycle of a sunflower? Look at the diagrams carefully and use words from the diagram to help form your answer. Write your response to the question on the lines below.

29. Water is stored underground, as well as in oceans and ice caps. Such underground storage reservoirs are called:

a. Storage tanks
b. Aquifers
c. Evaporators
d. Runoff

30. An ice cube tray is filled with water. The tray is placed in a freezer for 12 hours. The water freezes forming ice. The tray is removed from the freezer and left on a counter for 12 hours. What changes as the water becomes ice and then melts back into water?

a. The amount of energy in the water
b. The amount of water in the ice cube tray
c. The number of water molecules
d. The mass of the water molecules

31. Several people were asked to jump on a large trampoline, one at a time, to see who could jump the highest. The results were recorded in the data table below.

Person	Weight	Highest Jump
Sarah	59 kg	1.35 m
Mark	75 kg	1.81 m
Isaac	78 kg	1.93 m
Valerie	64 kg	1.47 m
Crystal	66 kg	1.57 m

Based on the data, what can be concluded about the relationship between weight and jumping height?

a. The more a person weighs, the higher he or she could jump
b. The less a person weighs, the higher he or she could jump
c. A person's weight does not affect how high he or she was able to jump
d. The taller the person is, the higher he or she could jump

32. Which statement describes how Earth's relationship to the Sun allows the polar ice caps to remain frozen?

a. The equator receives the most direct sunlight while the poles receive the least amount of sunlight
b. The Moon reflects the Sun's light away from the North and South Pole preventing them from heating up
c. Wind currents constantly move air warmed by the Sun away from the North and South Pole and toward the equator
d. The Sun does not shine on the North or South Pole so they never warm up

33. Which option would result in a rapid change to the Earth's surface?

a. Volcanic eruption
b. Sediment deposits
c. Rapidly flowing river
d. Windy day

34. Which piece of safety equipment is designed to protect your eyes?

a. Glasses
b. Goggles
c. Gloves
d. Eye wash

35. How is force being applied to the box below?

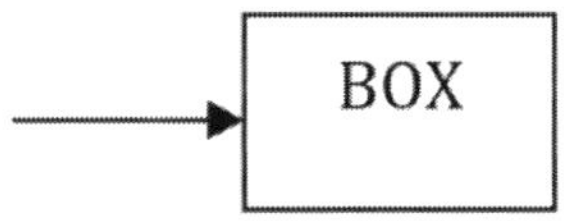

a. The box is being pulled forward
b. The box is being pushed forward
c. Gravity is forcing the box to move forward
d. Friction is forcing the box to stop

36. Examine the circuits below.

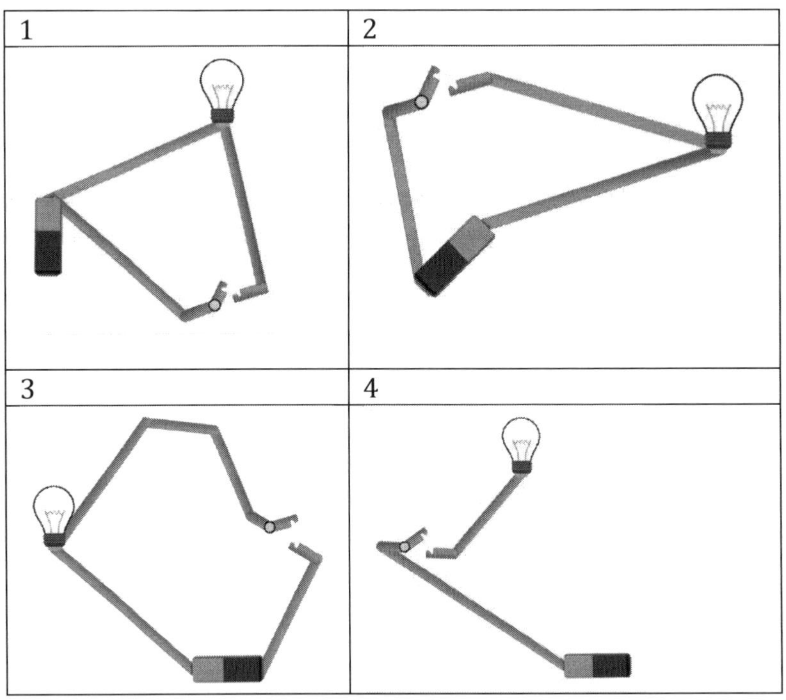

Which of these circuits will produce light when the switch is closed?

a. All of the circuits will produce light when the switch is closed.
b. Circuit 2 is the only circuit that will produce light when the switch is closed.
c. Circuits 3 and 4 are the only circuits that will produce light when the switch is closed.
d. Circuits 2 and 3 are the only circuits that will produce light when the switch is closed.

37. Which of the following does not represent a physical change?

a. Salt dissolved in water
b. A spoiling apple
c. Sand dissolved in water
d. Pulverized rock

38. Polar bears, seals and walrus' are all arctic animals that have a thick layer of blubber or fat beneath their skin. What purpose does this layer of blubber serve?

a. Protects them from predators
b. Helps to preserve body heat
c. Aids in finding food
d. Prevents them from drowning in water

39. The school uses a bell system to notify students that class is about to start. Which of the following uses the same type of energy as the bell system to accomplish its purpose?

a. The ovens in the cafeteria
b. The fire alarm
c. The lamps in the library
d. A handheld pencil sharpener

40. Why do some cacti have folds in their outer skin?

a. The folds can expand, allowing the cacti to absorb and store water when it rains
b. The folds help protect the cacti from predators
c. The folds allow them to absorb nutrients from the air
d. The folds provide support, allowing the cacti to grow tall

41. The factor most responsible for determining which season an area is experiencing is

a. Elevation
b. The position of the earth in its orbit around the sun
c. Latitude
d. Air masses

42. What object is needed at the question mark so that when switch 2 is closed, both buzzers will sound? Why is this object needed? Write your response to these questions on the lines below.

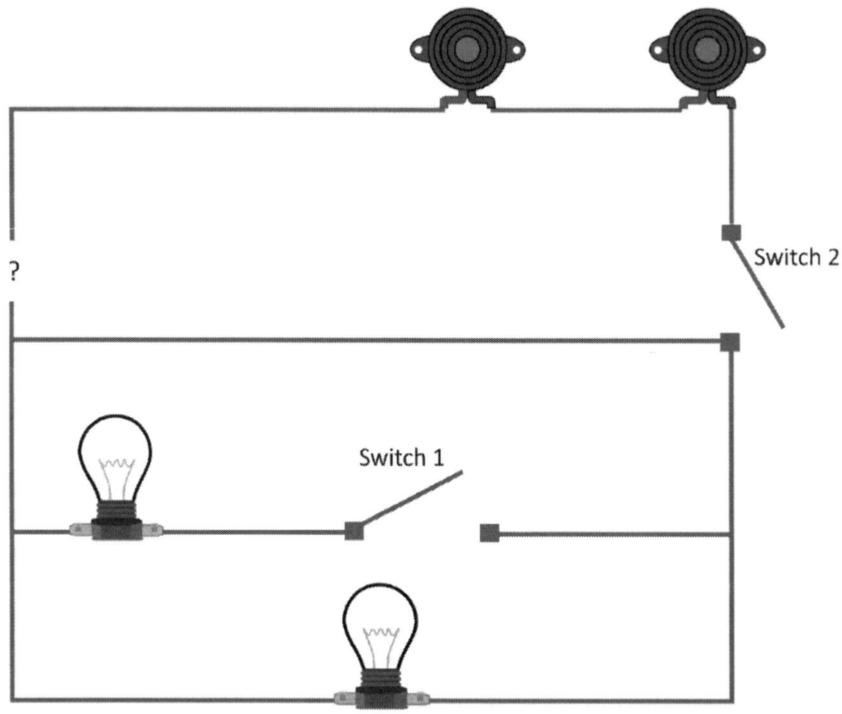

43. Students were studying the life cycles of apple trees. They learned that after a seed sprouts and an apple tree begins to grow, it takes five to eight years for the tree to actually start producing apples.

What would happen if apples suddenly stopped producing seeds?

a. Apple trees would continue to grow, but they would stop producing apples.
b. The apple tree would start producing a different type of apple.
c. Apple trees would eventually become extinct because an important part of the life cycle was taken away.
d. The apple tree would make the necessary adaptations and continue to produce apples without the seed part of the life cycle.

44. Which of the following processes is NOT part of the formation of sedimentary rock?

a. Layering
b. Cementation
c. Compaction
d. Heat

45. Which of the following is correct regarding how objects produce sound and how we hear it?

a. Sound is something that causes vibrations in the air.
b. Sound waves are vibrations disturbing media lie air.
c. Sound causes vibrations in the air but not in the ear.
d. Sound is the same energy type in the ear and brain.

46. The teacher asked two students to help her move a heavy box to the classroom across the hall in the direction of the arrow.

Which of the following describes the easiest way for the students to move the box?

a. Student 1 should take the rope and pull the box in the direction of the arrow. Student 2 should stand on the opposite side of the box and push it in the direction of the arrow.
b. Student 1 should take the rope and pull the box in the opposite direction of the arrow. Student 2 should stand on the same side of the box as student 1 and push the box in the opposite direction of the arrow.
c. Student 1 and student 2 should both stand on the opposite side of the box from the rope and take turns pushing the box in the direction of the arrow.
d. Student 1 should take the rope and pull the box in the direction of the arrow. Student 2 should stand on the opposite side of the box and pull it in the opposite direction of the arrow.

47. What is the main way that water from the ocean becomes incorporated into the water cycle?

a. Warm water currents heat up until the water within them evaporates.
b. The Sun heats the water on the surface of the ocean until it evaporates.
c. The thermal energy from marine life heats the water until it evaporates.
d. Underwater volcanoes erupt and cause the water above to evaporate.

48. Which of the following observations provides the best evidence that sound can travel through solid objects?

a. Sound waves cannot travel through a vacuum.
b. The atoms of a solid are packed tightly together.
c. If you knock on a solid object, it makes sound.
d. You can hear a sound of the other side of a solid wall.

49 A man is helping his daughter on the swings in their backyard.

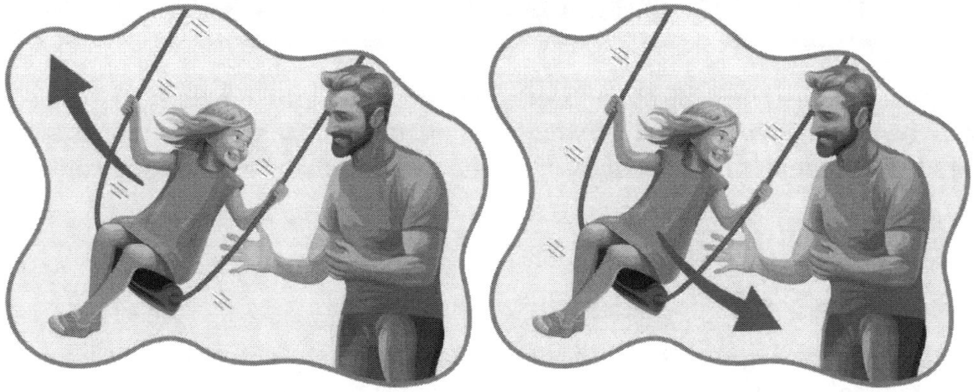

What force causes the girl to swing higher as she moves forward on the swing? What force will cause the girl to fall backwards on the swing when the man steps out of the way? Write your response to these questions on the lines below.

____

____

____

____

____

50. A fifth-grade science student completed a lab that required mixing 15 grams of salt into 500 milliliters of water. The student wrote the following observations in her lab journal.

1. Water is a liquid and takes the shape of its container.
2. Salt is no longer present once it is mixed into the water.
3. The mass of the salt does not change throughout the entire experiment.
4. Salt dissolved in water is an example of a solution.

Which observation is incorrect?

a. Observation 1
b. Observation 2
c. Observation 3
d. Observation 4

Answer Key and Explanations for Test #2

1. A: In the ecosystem described, birds eat beetles, crickets, and grasshoppers. If a pesticide were to kill off part of the beetle population, birds would be forced to consume more crickets and grasshoppers to meet their energy needs. As a result, there would be a decrease in the cricket population.

2. B: Herbivores are organisms that eat producers (plants) only. In the ecosystem described, both grasshoppers and crickets are herbivores, eating corn and grass.

3. A: Animals breathe in oxygen and breathe out carbon dioxide in a process called respiration. Plants use the carbon dioxide that animals exhale, in the process of photosynthesis, and convert it into oxygen.

4. A: Earth sits on its axis at a tilt of 23.4°. Choice B would result in Earth rotating sideways and choice C is straight up and down. Choice D is not even physically possible.

5. A: A hypothesis is an educated guess formulated by the experimenter based on prior experiments and research. A hypothesis includes what the experimenters think is going to happen and their reasoning.

6. B: Tides are caused by the effect of the Moon's gravity and the Sun's gravity pulling on the ocean waters.

7. C: The incorrect statement in the paragraph is "it takes about 48 hours for the Earth to rotate one full time around its axis." It takes approximately 24 hours for the Earth to rotate one full time on its axis. It is correct that the Sun does not move (A), the Earth spins on its axis (B), and the Earth revolves around the Sun (D).

8. A: A food chain shows how energy flows from one consumer to another. The arrows point in the direction that energy moves. For example, if an arrow points from a flower to a bee, then the energy from the flower flows to the bee as the bee eats the nectar of the flower.

9. A: Lions live in a wide open area where there are few large objects to hide behind. In order to get close enough to their prey to chase and attack it, they must be able to sneak up on it. Their coloration is similar to the color of the tall grasses where they live. This allows the lions to blend into their surroundings and get close to their prey.

10. B: Iron is a conductor of heat so the handle on the skillet will get hot quickly. Rubber is an insulator. The rubber sleeve will protect the user from getting burned by touching the hot iron handle of the skillet, so choice B is correct. Rubber is less dense than water, and by itself, the rubber sleeve would float in the water, so option A is incorrect. Options C and D are incorrect because both state that rubber is a good conductor, while rubber is actually an insulator.

11. C: Wind farms harvest energy from wind to use in combination with or in place of fossil fuels. The biggest limiting factors for wind farms are the inconsistency and unpredictability of wind and the difficulties involved in storing the energy that has been harvested. Wind farms can only produce as much energy as they receive from wind.

12. B: Carnivores eat meat only. The only choice that is not meat is acorns, which are plant products.

13. C: In this case, the term "how long" refers to the time it takes to do something, such as how long it takes a boy to run 100 meters. The instrument to measure time is a stopwatch.

14. D: The iron, toaster, and hair dryer all convert electrical energy to thermal energy to accomplish their main functions (D). They use electrical energy for a power source and thermal energy to heat clothes, food, and hair. While some of the items might make sounds (A), have lights (C), or use mechanical energy (B), these are not required for the objects to complete the tasks they were designed for.

15. B: A cooking pot must be able to conduct heat from the stove into the food being prepared. The only material that is a conductor from the choices given is copper.

16. C: Flowers, trees, and shrubs are all plants, which are producers. Producers are able to make their own food using energy from the Sun in a process known as photosynthesis.

17. B: A mirror is a reflective surface. The flat shiny surface reflects any incoming light allowing a person standing in front of the mirror to see their reflection.

18. A: Tropical environments typically receive rain 5 to 6 times per week, even if it is only a brief afternoon shower. In its new environment the tropical plant is not receiving enough water to survive.

19. 8: Mercury, Venus, Earth, Mars, Jupiter, Saturn, Uranus, and Neptune.

20. C: Since the dinosaur skeleton shows that the dinosaur had wings, the student would be able to determine that the dinosaur probably got from one place to another by flying. The skeleton does not show the dinosaur's teeth, so it is impossible to know what it ate (A). The fossil does not show the time period in which the dinosaur lived (B) or if it traveled in a pack or alone (D).

21. B: Algae are a type of plant that grows in water. Because algae are plants they are able to capture light from the Sun and transform it into energy during the process of photosynthesis.

22. B: A variable is something that is changed in the experiment. It is being tested against other similar things. In this experiment, the variable is the type of window cleaner.

23. B: The empty shopping cart does not weigh very much and is easy to push. As groceries are added to the empty cart, the cart gains mass. By the time the shopping cart is full, it has more mass than it began with and requires more force to push it.

24. D: The table states that a person's mass remains the same, but weight will fluctuate with the force of gravity. It also states that the more gravity acting on an object, the more the object will weigh. Neptune has a greater gravitational force than Earth, so the person's weight would increase.

25. C: When the bat makes contact with the ball, the bat will push the ball forward and away from the batter. The bat provides the force that interacts with the ball, and the force the bat uses is a push (D). The force of gravity will eventually pull the ball to the ground (A). However, because the ball was thrown using a force that sends it in a different direction, it takes longer for gravity to pull it down. Additionally, the bat changes the direction of the ball using another force, so it takes even more time for gravity to pull the ball to the ground. The ball would not go straight up (B) or straight down because the push of the bat forces the ball to move straight forward.

26. C: The table shows that the largest dogs have the lowest pitch barks ranging from 8 to 12 hertz. The smaller dogs, such as the Chihuahua and spaniel, have higher pitch barks of 17 and 14 hertz.

27. A: The Moon does not emit any light of its own. However, the Moon does reflect light from the Sun. The phase of the Moon that we are able to see on Earth, such as full moon, crescent, or no moon, depends on where the Moon is positioned in relation to the Sun and Earth.

28. Answers may vary: Possible similarities could include:

1. The frog produces many eggs, and the sunflower produces many seeds.
2. The tadpole grows legs in the froglet stage, and the baby plant grows a flower in the adolescent stage.
3. The adult frog produces offspring, and the adult sunflower produces offspring.
4. The adult frog and the adult sunflower produce many offspring at one time.

29. B: An aquifer (a word derived from the Latin root *aqua*, meaning water) is any geologic formation containing ground water.

30. A: When water changes from a liquid to a solid the amount of thermal energy is decreasing. The molecules move closer together, which creates the solid crystal structure of the ice. As the ice melts back into a liquid, thermal energy level increases, causing the molecules to spread out again.

31. A: The table shows a clear correlation between mass and jump height. If you put the list of jumpers in order from highest mass to lowest mass, they will also be in order from highest jump height to lowest jump height. Choice D can be eliminated immediately because there is no information about any of the individuals' heights in the data table or in the question.

32. A: The equator receives the most direct sunlight while the North and South Pole receive the least amount of direct sunlight. In addition, the atmosphere helps to reflect some of the Sun's heat energy away from the poles. Finally, the reflectivity of the ice helps to deflect some of the heat away from the polar ice caps.

33. A: A volcanic eruption result in a rapid change to the Earth's surface. A volcanic eruption could cause rapid change because the magma released by the volcano could create new rock formations or mountains when it hardens.

34. B: Goggles are proper protective eyewear for laboratory. Glasses provide almost no protection as they could break and they are open around the sides where chemicals could splash into the eyes. An eye wash is used to rinse out eyes if chemicals accidently get splashed into them.

35. B: The picture shows an arrow behind the box pushing it forward. If the box were being pulled, the arrow would be in front of the box pulling the box forward. Gravity pulls downward, not forward.

36. D: A circuit is complete when it contains an energy source, an electrical device, and conductive wire that connects the source and device together in a complete loop. The entire circuit must be closed. Only circuits 2 and 3 are constructed correctly (D). Circuit 1 would not produce light because the conducting wires are not connected to one end of the energy source, the battery (A). Circuit 4 would not produce light because there are no wires connecting one side of the energy source, the battery, to the other side of the device that is using the energy, the light bulb (C).

37. B: A spoiling apple has undergone a chemical change (one substance is changed into another). Dissolving both sand and salt in water, refers to a physical change, since the salt and water and the sand and water can be separated again by evaporating the water, which is a physical change.

Pulverized rock is also an example of a physical change where the form has changed but not the substance itself.

38. B: Polar bears, seals, and walrus' all live in arctic climates where it is very cold all year round. The thick layer of blubber that these animals have below the surface of their skin helps them preserve body heat keeping them warm in the frigid environment.

39. B: The bell system in the school uses sound energy to let students know class is about to start. The fire alarm also uses sound energy to alert students to the possibility of a fire (B). The ovens use electrical and thermal energy to heat food (A), the lamps use electrical and light energy to provide light (C), and a handheld pencil sharpener uses mechanical energy to sharpen pencils (D).

40. A: Cacti live in a desert environment where water is scarce. When it rains, cacti must be able to absorb and store as much water as possible so that they have water available during the dry periods. The folds on the outside of a cactus allow it to expand and fill with water. After a heavy rain, cacti will appear round and plump because the folds are full of water.

41. B: As the earth travels around the sun, it tilts towards the sun or away from it, depending on where it is in its orbit. This tilt is what determines the seasons. The earth is tilted towards the sun in summer and away from it in winter. Elevation a. and latitude c. of an area remain constant throughout the year, so they cannot account for the changing of the seasons. Air masses d. can influence the weather, but don't determine the seasons.

42. Answers may vary: A correct response to this question will state that a power source, such as a battery, is needed at the question mark so that the buzzers will sound when switch 2 is closed. A correct response will also state that all circuits must have a power source that provides the energy needed to power the circuit. Complete circuits must include a power source and a device, and they must be connected by conductive wires to form a complete loop.

43. C: If the apples suddenly stopped producing seeds, no new apple trees could grow. The seed is the first step in the life cycle of an apple tree, and without the seed, no new apple trees could grow. Eventually apple trees would become extinct. The apple tree is what grows from the seed, so without the seed, there could be no apple trees nor apples of any kind. Trees can adapt to their environments, but it is impossible to grow an apple tree without a seed.

44. D: The formation of sedimentary rock does not include heat. Of the three types of rock igneous, sedimentary and metamorphic, heat is essential to two: igneous and metamorphic. Sedimentary rocks are formed by sediments that get deposited and then compacted or cemented together. Sedimentary rocks are classified into detrital, organic or chemical sediments. Answer A, layering, is correct since sediments can be deposited or otherwise formed in layers. Answer B, cementation, is also called lithification. Answer C, compaction, refers to the pressure forming sedimentary rock leading to cementation.

45. B: Vibration is caused by anything physical moving rapidly back and forth. When an object vibrates, the vibrations disturb the medium surrounding it, which may be solid matter, liquids, or gases like the air. Sound does not cause vibrations in the air (A) but vice versa: vibrations are a type of energy. The sound caused by vibration is acoustic energy, and vibratory movements disturb their medium, like the air. Sound causes vibrations not only in the air, but also in the ear (C) when they reach it. Whereas sound waves traveling through the air and through the outer and middle ear are acoustic energy, once they reach the inner ear, it converts that acoustic energy into electrical energy that is sent via nerves to the brain (D).

46. A: The greatest force that pushes the box in the direction of the arrow is created when student 1 pulls the rope in the direction of the arrow while student 2 stands on the opposite side of the box and pushes it in the direction of the arrow. In option B, the box would move in the wrong direction. In option C, only one student is pushing at a time, so this would not result in the greatest possible force. In option D, the students are pulling in opposite directions, so the force they create may not send the box in the correct direction.

47. B: The Sun shines down and heats the water on the surface of the oceans. This causes ocean water to evaporate and follow the steps of the water cycle. Water in warm water currents does not get hot enough to evaporate, so choice A is incorrect. Animals that live in the ocean cannot produce enough heat to cause water to evaporate, so choice C is incorrect. While underwater volcanoes may erupt and drastically heat up water, this is not the way that most water vapor from the ocean is created. Choice D is incorrect.

48. D: Sound cannot travel through a vacuum, though it doesn't necessarily suggest that it *can* travel through solids. Nor does the fact that atoms are packed tightly together demonstrate the fact that sound can travel through a solid. The fact that a sound is produced by knocking on a solid object also does not prove sound can pass through the object. However, if you hear a sound on the other side of a solid wall, the sound must have traveled through the wall.

49. Answers may vary: An outside force must be involved for movement of the girl on the swing to occur. The force of the man pushing his daughter moves her higher and forward on the swing. Since gravity only pulls things toward the Earth's center, gravity is the force that is acting on the girl as she falls backwards on the swing when the man moves out of the way.

50. B: Observation 2 (B) is incorrect. Salt is still present in the water. It has dissolved, so it is not clearly visible, but it is still in the water. Observation 1 (A) is correct: water is a liquid and takes the shape of its container. Observation 3 (C) is correct: when a substance is added to a liquid, the mass of the substance does not change. Observation 4 (D) is correct: dissolving salt in water is an example of creating a solution.

Online Resources

Due to our efforts to try to keep this book to a manageable length, we've created a link that will give you access to all of your online resources:

mometrix.com/resources719/vsolg5sci

It's Your Moment, Let's Celebrate It!

Share your story @mometrixtestpreparation